ATP 3-05.40 (FM 3-05.140)

Special Operations Sustainment

May 2013

DISTRIBUTION RESTRICTION: Distribution authorized to U.S. Government agencies and their contractors only to protect technical or operational information from automatic dissemination under the International Exchange Program or by other means. This determination was made on 15 November 2012. Other requests for this document must be referred to Commander, United States Army John F. Kennedy Special Warfare Center and School, ATTN: AOJK-CDI-CID, 3004 Ardennes Street, Stop A, Fort Bragg, NC 28310-9610.

DESTRUCTION NOTICE: Destroy by any method that will prevent disclosure of contents or reconstruction of the document.

FOREIGN DISCLOSURE RESTRICTION (FD 6): This publication has been reviewed by the product developers in coordination with the United States Army John F. Kennedy Special Warfare Center and School foreign disclosure authority. This product is releasable to students from foreign countries on a case-by-case basis only.

Headquarters, Department of the Army

This publication is available at Army Knowledge Online (https://armypubs.us.army.mil/doctrine/index.html).
To receive publishing updates, please subscribe at http://www.apd.army.mil/AdminPubs/new_subscribe.asp.

*ATP 3-05.40 (FM 3-05.140)

Army Techniques Publication
No. 3-05.40

Headquarters
Department of the Army
Washington, DC, 3 May 2013

Special Operations Sustainment

Contents

		Page
	PREFACE	v
	INTRODUCTION	vi
Chapter 1	OVERVIEW OF SUSTAINMENT	1-1
	Army Sustainment Structures	1-1
	Principles of Sustainment	1-5
	Sustainment Warfighting Function	1-7
	Expeditionary Logistics Imperatives	1-9
Chapter 2	ARMY SPECIAL OPERATIONS FORCES LOGISTICS SUPPORT FRAMEWORK	2-1
	Army Support Structure	2-1
	Logistics Integration Into Operations	2-2
	Support Relationships	2-2
	Logistics Planning and Preparation	2-3
	Army Force Generation	2-5
	Special Operations Force Generation	2-5
	Geographic Combatant Commanders' Theater Logistics Environment	2-8
	Logistics Information Systems Support	2-10

Distribution Restriction: Distribution authorized to U.S. Government agencies and their contractors only to protect technical or operational information from automatic dissemination under the International Exchange Program or by other means. This determination was made on 15 November 2012. Other requests for this document must be referred to Commander, United States Army John F. Kennedy Special Warfare Center and School, ATTN: AOJK-CDI-CID, 3004 Ardennes Street, Stop A, Fort Bragg, NC 28310-9610.

Destruction Notice: Destroy by any method that will prevent disclosure of contents or reconstruction of the document.

Foreign Disclosure Restriction (FD 6): This publication has been reviewed by the product developers in coordination with the United States Army John F. Kennedy Special Warfare Center and School foreign disclosure authority. This product is releasable to students from foreign countries on a case-by-case basis only.

*This publication supersedes FM 3-05.140, 12 February 2009.

Contents

Chapter 3	SUSTAINMENT BRIGADE	3-1
	Staff Structure	3-1
	Mission	3-1
	Concept of Employment	3-2
	Organization	3-3
Chapter 4	SPECIAL FORCES GROUP	4-1
	Overview of Special Forces Sustainment	4-1
	Group Support Battalion	4-1
	Special Forces Battalion	4-7
	Unconventional Warfare Sustainment	4-14
	Foreign Internal Defense Sustainment	4-16
Chapter 5	RANGER REGIMENT	5-1
	Mission	5-1
	Organization	5-1
	Ranger Support Operations Detachment	5-2
	Ranger Support Company	5-4
	Headquarters Detachment	5-6
	Distribution Platoon	5-6
	Sustainment Platoon	5-7
	Maintenance Platoon	5-8
	Base Support Section	5-8
	Chemical, Biological, Radiological, and Nuclear Section	5-9
	Ranger Logistics Support	5-9
	Anniston Army Depot Contingency Stocks	5-9
	Army Health System Support	5-9
	Tactical Medical Evacuation	5-10
	Religious Support	5-10
Chapter 6	ARMY SPECIAL OPERATIONS AVIATION COMMAND	6-1
	Mission	6-1
	Planning Considerations	6-3
	Contingency Planning	6-3
	Crisis Action Planning	6-3
	Logistics Support	6-4
	Health Service Support/Force Health Protection	6-7
	Funding and Finance Support	6-7
	Engineer Support	6-8
	Forward Arming and Refueling Point Operations	6-8
	Logistics in Developed and Undeveloped Theater of Operations	6-8
	Contracting	6-9
Chapter 7	CIVIL AFFAIRS BRIGADE	7-1
	Mission	7-1
	Organization	7-1
	Logistics Support	7-1
	Planning and Preparation Considerations	7-3
	Statement of Requirements	7-3

Contents

	Army Health System Support	7-3
	Stability Operations	7-4
	Major Operations	7-4
Chapter 8	**MILITARY INFORMATION SUPPORT OPERATIONS COMMAND**	**8-1**
	Mission	8-1
	Organization	8-1
	Sustainment	8-2
	Statement of Requirements	8-4
	Special Maintenance Considerations	8-4
	Army Health System Support	8-5
Chapter 9	**ARMY HEALTH SYSTEM SUPPORT**	**9-1**
	Army Special Operations Forces Medical Considerations	9-1
	Health Service Support	9-3
	Force Health Protection	9-8
Chapter 10	**PERSONNEL SERVICES SUPPORT**	**10-1**
	Human Resources Support	10-1
	Financial Management	10-3
	Legal Support	10-7
	Religious Support	10-9
	Band Support	10-9
Chapter 11	**CONTRACTING AND HOST-NATION SUPPORT**	**11-1**
	Types of Contractor Support	11-1
	Planning Contract Support	11-2
	Key Contracting Officials	11-2
Appendix A	LOGISTICS PLANNING CHECKLIST	A-1
Appendix B	JOINT OPERATIONAL STOCKS	B-1
Appendix C	CLASSES AND SUBCLASSES OF SUPPLY	C-1
Appendix D	SITE SURVEY CHECKLIST	D-1
Appendix E	STATEMENT OF REQUIREMENTS FORMAT	E-1
Appendix F	HEALTH THREAT AND MEDICAL INTELLIGENCE	F-1
Appendix G	CONSIDERATIONS IN PLANNING MEDICAL EVACUATIONS	G-1
	GLOSSARY	Glossary-1
	REFERENCES	References-1
	INDEX	Index-1

Figures

Figure 2-1. Typical ARSOF sustainment structure ... 2-3
Figure 3-1. ARSOF tactical-to-strategic planning model ... 3-2
Figure 3-2. 528th Sustainment Brigade (Special Operations) (Airborne) ... 3-3

Figure 3-3. Example of ASPO cell deployments... 3-6
Figure 4-1. Group support battalion 4-3
Figure 5-1. 75th Ranger Regiment (Airborne) organization 5-1
Figure 5-2. Ranger Special Troops Battalion organization... 5-2
Figure 5-3. Ranger Regiment headquarters and headquarters company organization........ 5-3
Figure 5-4. Ranger battalion organization 5-4
Figure 5-5. Ranger Support Company organization... 5-5
Figure 6-1. Army Special Operations Aviation Command 6-2
Figure 6-2. 160th Special Operations Aviation Regiment (Airborne) organization.............. 6-2
Figure 7-1. Active Army Civil Affairs (Airborne) organization 7-2
Figure 8-1. Military Information Support Operations Command (Airborne) organization 8-2
Figure 9-1. ARSOF health service support capabilities.. 9-5
Figure 9-2. Organic and nonorganic Army Health System support.................................. 9-9
Figure A-1. Logistics planning checklist... A-1
Figure C-1. Description of classes and subclasses of supplies C-1
Figure D-1. Sample site survey checklist... D-1
Figure E-1. Statement of requirement format E-1
Figure F-1. Medical intelligence preparation of the battlefield template F-2
Figure F-2. Medical intelligence support appendix format... F-7

Tables

Table 4-1. Organic and nonorganic health service support/force health protection............ 4-9
Table 6-1. Water requirements for aircraft washing and engine flushing (gallons).............. 6-5
Table 6-2. Medical personnel authorizations for the special operations aviation regiment.. 6-7

Preface

Army Techniques Publication (ATP) 3-05.40 provides the United States (U.S.) Army special operations forces (ARSOF) commander and staff information on the structure and functions involved in sustainment activities. The acronym ARSOF represents Civil Affairs (CA), Military Information Support operations (MISO), Rangers, Special Forces (SF), Special Mission Units, and Army special operations aviation (SOA) forces assigned to the United States Army Special Operations Command (USASOC)—all supported by the Sustainment Brigade (Special Operations) (Airborne) (SB[SO][A]).

The principal audience for ATP 3-05.40 is ARSOF, joint, and land component force commanders and staff; the publication provides a broad understanding of ARSOF sustainment. This publication also provides guidance for ARSOF commanders who determine the force structure, budget, training, materiel, and operations and sustainment requirements necessary to prepare ARSOF to conduct their missions. This Service doctrine is consistent with joint doctrine.

Commanders, staffs, and subordinates ensure their decisions and actions comply with applicable U.S., international, and, in some cases, host nation laws and regulations. Commanders at all levels ensure their Soldiers operate in accordance with the law of war and the rules of engagement. (See Field Manual [FM] 27-10, *The Law of Land Warfare*.)

ATP 3-05.40 uses joint terms where applicable. Selected joint and Army terms and definitions appear in both the glossary and the text.

This publication reflects and supports the Army sustainment doctrine as stated in Army Doctrine Reference Publication (ADRP) 4-0, *Sustainment*, yet provides unique techniques, specific to the audience of this publication, that are not covered in ADRP 4-0. This publication is not intended as a stand-alone reference; rather, it is intended to be used in conjunction with existing doctrine. Examples and graphics are provided to illustrate principles and doctrine—not to serve as prescriptive responses to tactical situations. The appropriate leaders on the ground make the final decision for the best way to support their forces and defend their logistics units.

ATP 3-05.40 applies to the Active Army, Army National Guard (ARNG)/Army National Guard of the United States (ARNGUS), and the United States Army Reserve (USAR) unless otherwise stated.

This publication is unclassified to ensure Armywide dissemination and to facilitate the integration of special operations sustainment in the preparation and execution of campaigns and major operations. Unless this publication states otherwise, masculine nouns and pronouns do not refer exclusively to men. The proponent of this publication is the United States Army John F. Kennedy Special Warfare Center and School (USAJFKSWCS). The preparing agency is the Joint and Army Doctrine Integration Division, Army Special Operations Capabilities Integration Center, Capabilities Development and Integration Directorate, USAJFKSWCS. Submit comments and recommended changes on DA Form 2028 (Recommended Changes to Publications and Blank Forms) to Commander, USAJFKSWCS, ATTN: AOJK-CDI-CID, 3004 Ardennes Street, Stop A, Fort Bragg, NC 28310-9610; by e-mail to JAComments@ahqb.soc.mil; or submit an electronic DA Form 2028.

Introduction

ATP 3-05.40 is a revision of FM 3-05.140, *Army Special Operations Forces Logistics*, last published in 2009; it provides updates to sustainment requirements and infrastructure within the special operations (SO) community. The publication provides doctrinal guidance on the organization and capabilities of support units for SO. It outlines the necessary requirements for conducting, planning, preparing, executing, and assessing ARSOF sustainment. ATP 3-05.40 has eleven chapters, which are summarized in the following paragraphs.

Chapter 1 discusses that, to be effective, Army and ARSOF planners must understand the ARSOF sustainment organizations' operational concepts, the basic principles of sustainment, sustainment warfighting functions, and the ARSOF expeditionary logistics imperatives. The publication provides a clear understanding that ARSOF sustainment is not self-sufficient and that it is reliant upon regional or combatant command theater of operations infrastructure for virtually all support above organic unit capabilities.

Chapter 2 discusses how USASOC has aligned its ARSOF sustainment organizations and activities in support of the U.S. Army's concept of modularity and force projection—an alignment allowing ARSOF to integrate organic support elements within the theater of operations support structure for responsive and continuous sustainment of ARSOF units.

Chapter 3 introduces the 528th SB(SO)(A) and how it is unique when compared to other Army sustainment brigades in that it maintains global situational awareness of sustainment support structures, is focused at the operational level for sustainment planning and synchronization, and is designed to deploy small, modular teams.

Chapter 4 introduces the organic sustainment capabilities of the Special Forces group (airborne) (SFG[A]). It discusses how SF units are less than comparable to Army units by design and how SF units are reliant upon regional and geographic theater of operations support capabilities for support beyond their organic capabilities.

Chapter 5 introduces the Ranger support operations detachment (RSOD) and the organic Ranger support companies (RSCs) enhancing the expeditionary capabilities of the Ranger Regiment.

Chapter 6 discusses the unique support requirements of the Army Special Operations Aviation Command (ARSOAC). It explains how conventional force logistics organizations and procedures are normally adequate for SOA requirements and how standard procedures are in place to handle the few SOF-unique requirements through the theater special operations command (TSOC) and the Army special operations forces liaison element (ALE).

Chapter 7 discusses how the type of operation, deployment sequence, unit-basing, and area of responsibility (AOR) shape the logistics environment for CA forces. It also discusses how geographic theater sustainment command (TSC) organizations and procedures are normally adequate for CA requirements and how the 528th SB(SO)(A) procedures are in place to handle the few CA-peculiar equipment requirements.

Chapter 8 discusses the challenges of the planner to develop a support concept for a joint Military Information Support (MIS) task force that operates from multiple bases, ranging from the continental United States (CONUS) through the communications zone to the joint operations area (JOA). The task becomes more complex by the requirement to derive common support from ARSOF, joint, and multinational sources.

Chapter 9 addresses the Army Health System (AHS), which consists of health service support (HSS), a component of the sustainment warfighting function, and force health protection (FHP), a component of the protection warfighting function. These two components, though separated into different warfighting functions, must be considered equally during SO planning and execution.

Chapter 10 discusses the limitations of personnel services support within ARSOF and how ARSOF units are reliant upon the geographic combatant commander (GCC) within the theater of operations for personnel services support.

Chapter 11 discusses how ARSOF units have no organic operational contracting capability and are reliant upon contracting support from the theater GCC, Service components, and USASOC or United States Special Operations Command (USSOCOM) contracting activities.

This page intentionally left blank.

Chapter 1
Overview of Sustainment

For the Army, sustainment is the provision of logistics, personnel services, and health services support necessary to maintain operations until successful mission completion (Army Doctrine Publication [ADP] 4-0, *Sustainment*). ARSOF sustainment is not self-sufficient; it is reliant upon regional or combatant command theater of operations infrastructure for virtually all support above unit organic capabilities. The planning and execution of support to ARSOF must be nested within the combatant commander's (CCDR's) concepts of operation and support, as well as tailored to interface with the theater of operations logistics structures. To be effective, Army and ARSOF planners must understand the ARSOF sustainment organizations' operational concepts, the basic principles of sustainment, sustainment warfighting functions, and the ARSOF expeditionary logistics imperatives.

ARMY SUSTAINMENT STRUCTURES

1-1. ARSOF sustainment structures are lean and unable to provide all sustainment functions required to support ARSOF missions. ARSOF sustainment structures are designed to perform the following tasks:
- Enable expeditionary ARSOF missions.
- Deploy early and rapidly.
- Collocate and habitually train with the supported unit.
- Fill immediate and critical logistical requirements with organic formations.
- Provide the capability to plug into theater of operations logistics structures, therefore achieving required logistics staying power.
- Tie the Army SO units to the operational theater of operations support structure.

1-2. Only those Army SO units designed for mission command of tactical SO, SFGs, and the Ranger Regiment are resourced with organic sustainment support capabilities. The SOA, CA, and MISO units possess only organizational-level sustainment personnel because they are designed to deploy and operate while task-organized under an ARSOF-led joint special operations task force (JSOTF), with an SFG, or with the Ranger Regiment from which they would receive direct support and sustainment. The USASOC also consists of the 528th SB(SO)(A), which has a global, operational-level focus. The unit's mission is to set the operational-level logistics conditions to enable expeditionary ARSOF missions within Army theater of operations sustainment infrastructures.

1-3. Army common-user logistics (CUL) is the responsibility of the theater of operations Army Service component commands (ASCCs) under Title 10, United States Code (10 USC), *Armed Forces*. Therefore, Army SO rely upon ASCC logistics structures to provide Service CUL to all Army forces in the area of operations (AO) regardless of command structure. ARSOF units lack the robust logistics structure normally associated with Army. ARSOF routinely arrive into the AO early, execute forced-entry operations, and operate independently in small teams. Because of these factors, ASCC logistics support to ARSOF must be tailored to meet logistics requirements based upon mission, enemy, terrain and weather, troops and support available, time available, and civil considerations (METT-TC). For example, an SFG-led JSOTF with its organic group support battalion (GSB) cannot simply plug into the distribution network of a single sustainment brigade and execute tactical distribution to each of the SF battalions, companies, and Special Forces operational detachments A (SFODAs) in its task organization. Most likely, a comprehensive concept of support, including multiple Army sustainment brigades and combat

Chapter 1

sustainment support battalions (CSSBs), spread across the joint operations area (JOA), will be required with some CSSBs delivering down to the SFODA level. In addition to Service CUL, ARSOF have requirements for SOF-peculiar equipment that requires supply, sustainment, and maintenance mechanisms outside of the Army-common support structure. SOF-peculiar sustainment requirements are the responsibility of USASOC and United States Special Operations Command (USSOCOM).

1-4. Operational-level logistics planning is critical not only to mission success but also to the ability of regional ASCCs to be responsive to SO sustainment requirements. ARSOF units operate under the command and control of theater special operations commands (TSOCs); therefore, operational-level logistics planning begins with the TSOC's joint concept of operations. The 528th SB(SO)(A) Army special operations forces liaison elements (ALEs) develop the corresponding operational-level ARSOF concept of support and coordinate logistics requirements with theater of operations ASCCs for resourcing SOF-peculiar requirements. These requirements are passed back through the 528th SB(SO)(A) to USASOC for resourcing. Refinement of the ARSOF concept of support for an operation is coordinated by the 528th SB(SO)(A), ARSOF commands, and executing units.

1-5. ARSOF sustainment units enable ARSOF missions by ensuring that operational-level logistics conditions are set through detailed planning prior to deployment. The 528th SB(SO)(A) focuses on operational-level logistics planning and synchronization (versus the tactical distribution focus of Army sustainment brigades).

528TH SUSTAINMENT BRIGADE (SPECIAL OPERATIONS) (AIRBORNE)

1-6. The 528th SB(SO)(A) is a deployable sustainment organization assigned to USASOC. The unit's mission is to set the operational-level sustainment conditions needed to enable ARSOF missions. Using forward-stationed ARSOF logistics elements and the modular and deployable Army special operations forces support operations (ASPO) cells, the brigade ensures sustainment requirements generated from operational plans developed at the TSOCs are integrated and synchronized with the ASCC support plan. Chapter 3 includes more in-depth information on the 528th SB(SO)(A).

SPECIAL FORCES GROUP (AIRBORNE)

1-7. Each SFG(A) possesses an organic GSB, with a subordinate sustainment and distribution company, maintenance company, medical company, and three forward support companies. Each SF battalion is supported by one of the GSB's forward support companies.

1-8. The GSB is a multifunctional, direct-support logistical organization organic to the SFG(A) with force structure and capabilities tailored to support. The GSB is a cornerstone of tactical ARSOF logistics formations. The GSB plans, coordinates, and executes logistical sustainment operations for the unit and, when directed, supports forces task-organized with the SFG(A), or an ARSOF-led JSOTF. Chapter 4 includes more in-depth information on the SFG(A).

75TH RANGER REGIMENT (AIRBORNE)

1-9. The 75th Ranger Regiment's mission is to plan and conduct SO against strategic and operational targets in pursuit of national or theater of operations objectives. The regiment consists of a regimental headquarters (HQ) with an RSOD, a Ranger Special Troops Battalion, and three Ranger battalions with organic RSCs. The Ranger Special Troops Battalion provides staff planning and supervision for all logistics within the regiment. The RSOD coordinates with logistics and force health protection (FHP) personnel in the areas of supply, maintenance, and movement management for the support of all units assigned or attached. The RSCs are multifunctional logistics companies that are organic to each Ranger battalion within the regiment and provide organizational and limited direct-support logistics. Chapter 5 includes more in-depth information on the 75th Ranger Regiment.

Overview of Sustainment

ARMY SPECIAL OPERATIONS AVIATION COMMAND

1-10. The ARSOAC mission is to equip, train, validate, conduct, and support special air operations by clandestinely penetrating hostile and denied airspace. The command deploys forces worldwide in support of contingency missions, the joint task force (JTF) commander, and the GCC. The command's battalions have organic centralized aviation unit maintenance and aviation intermediate maintenance capability for all assigned aircraft, armament, and avionics. However, they have very limited organic sustainment support capability and are dependent upon other ARSOF sustainment elements, the theater sustainment command (TSC), and the joint special operations air component. Chapter 6 includes more in-depth information on the ARSOAC.

MILITARY INFORMATION SUPPORT OPERATIONS COMMAND (AIRBORNE)

1-11. The Military Information Support Operations Command (Airborne) (MISOC[A]) assigned to USASOC consists of two ARSOF Military Information Support groups (MISGs). The MISO mission is to influence the behavior of foreign target audiences to support U.S. national objectives. MISO forces have limited unit-level sustainment capability and must establish logistics support relationships and identify contracting requirements early on (prior to deployment), especially when assigned to a JSOTF with ARSOF logistics elements, as in the GSB and RSC. Chapter 8 includes more in-depth information on the MISOC(A).

95TH CIVIL AFFAIRS BRIGADE (AIRBORNE)

1-12. The CA mission is to engage and influence the civil populace by planning, executing, and transitioning Civil Affairs operations (CAO) in Army, joint, interagency, and multinational operations. CA forces support commanders in engaging the civil component of their operational environment to enhance civil-military operations (CMO) or other stated U.S. objectives before, during, or after other military operations. ARSOF CA, like MISO and the 160th Special Operations Aviation Regiment (Airborne) (SOAR[A]), have only limited unit-level sustainment capability and must establish logistics support relationships early on (prior to deployment), especially when assigned to a JSOTF with ARSOF logistics elements, as in the GSB and RSC. Chapter 7 includes more in-depth information on the 95th Civil Affairs Brigade (Airborne).

HEADQUARTERS AND SUPPORTING STAFF ELEMENTS RESPONSIBLE FOR SUSTAINMENT

1-13. USSOCOM is a four-star combatant command that leads, plans, synchronizes, and, as directed, executes global operations against terrorist networks. USSOCOM has the Title 10 authority and responsibilities to train, organize, equip, and deploy combat-ready SOF to the combatant commands. USSOCOM has unique responsibilities in that it is not dependent upon the Services for its budget. Through the Major Force Program–11, USSOCOM is able to develop and acquire SO-peculiar equipment, material, supplies, and services. USASOC is the ARSOF component of USSOCOM and maintains command relationships with both Headquarters, Department of the Army (HQDA), and USSOCOM. The majority of ARSOF equipment and materiel is Army-common, making ARSOF dependent upon Army Major Force Program–2 funds for equipping and sustainment. Command guidance directs that USASOC leverage their Service component before going to USSOCOM for force structure, materiel solution, and sustainment.

1-14. The special operations acquisition and logistics (SOAL) center J-4 is a key staff element in support of ARSOF logistics. This staff section plans, coordinates, synchronizes, and integrates operational and strategic logistics and acquisition sustainment strategy in coordination with combatant commands, Services, components, and other agencies. The SOAL center is capable of providing rapid and focused acquisition and logistics support to SOF. The SOF support activity is capable of providing dedicated contractor logistics support capability, SOF aviation support and services, life-cycle sustainment management, and specified unit support for SO-peculiar equipment.

1-15. The USASOC, as an ASCC, is subject to logistics policies and programs from the office of the Army Deputy Chief of Staff (DCS) G-4 and other staff elements within HQDA. The Army G-4 is able to maintain domain-wide visibility over requirements, resources, and priorities, and acts with unity of effort in the planning and execution of logistics across the JOA. It also maintains operational awareness through daily updates in the Army operations center, through the HQDA G-3, Special Operations Division, and through Army database systems, such as the equipment common operational picture (ECOP) database.

1-16. The USASOC DCS G-4 mission is to plan, develop, manage, and execute the logistics policies, programs, and resources to sustain ARSOF in a joint environment in support of USSOCOM and CCDRs. Some key functions are to ensure logistics support to ARSOF mobilizing, deploying, and redeploying from exercises, operations, and contingencies. The G-4 plans and coordinates for theater of operations support and ammunition and transportation requirements, and monitors current operations to track sustainment of forces. It also performs contingency, crisis, mobilization, exercise, and reconstitution planning in coordination with USSOCOM, CCDRs, supporting subcombatant commands (theater of operations ASCCs), the Army staff, and TSOCs. The G-4 coordinates and, in some cases, manages sustainment programs required to support Army-common and SO-peculiar equipment assigned to ARSOF. The G-4 functions as the USASOC integrator for logistics Enterprise Information Systems and the standard Army management information system. Another function performed is property management, and a sensitive activities division oversees supply and maintenance requirements for special-purpose commodities.

1-17. The USASOC surgeon's office plays a vital role in AHS support. Even though it has limited organic capability and is dependent upon theater of operations support, its focus is ensuring the right organic and theater of operations medical support and services are being provided in a timely manner.

1-18. The USASOC contracting office plays a vital role in both the employment and sustainment of ARSOF. It plays a critical and integral role in obtaining supplies, services, and construction assistance in support of operations. In-theater, the expeditionary contracting command's contingency contracting battalion aligned with USASOC supports ARSOF task forces and their attached subordinate elements and units. For requirements that cannot be met by the contingency contracting battalion, the G-4 forwards requests to the TSOC for action.

1-19. The U.S. Army Special Forces Command DCS, G-4, is the principal staff element for sustainment matters. This element is responsible for addressing matters concerning the sustainment and sustainment readiness for the U.S. Special Forces Command. It performs tactical support planning and coordination, develops staff logistical estimates, and prepares the logistics and sustainment portion of command plans and orders. It serves as the focal point for logistics support to ARSOF mobilizing, deploying, and redeploying.

1-20. The Army's theater of operations-level ASCC HQ consolidates supporting functions currently executed by Army corps and SOF into a single operational command echelon directly supporting the CCDR. The ASCC's TSC executes operational logistics for support to the entire region, as well as Army, SO, joint, and multinational forces deployed to a JOA. The theater of operations ASCC reports directly to the Department of the Army serving as the Army's single point of contact for a combatant command or a functional component command. The ASCC performs Service-unique functions and tasks in support of the CCDR. In major combat operations, the ASCC commander may become the joint force land component commander and exercise operational control over tactical forces. The ASCC can also provide the HQ for a JTF in smaller-scale contingencies.

1-21. The mission of the TSC is to plan, prepare, rapidly deploy, and execute operational logistics within an assigned AO or JOA for the theater-level numbered ASCC or joint force commanders. The TSC provides single logistics mission command in theater of operations, simultaneously providing full-range support operations during deployment, employment, sustainment, and redeployment. The expeditionary sustainment command (ESC) provides forward-based mission command for logistics forces under the operational control of the TSC. The Army sustainment brigades are also subordinate commands of the TSC that plan, coordinate, synchronize, monitor, and control logistics operations. This single element provides mission command of the full range of logistics operations conducted at the operational theater of operations ASCC or tactical (corps/division/special operations task force [SOTF]) level. CSSBs are the building blocks of the Army and the 528th SB(SO)(A). They are modular and tailored to provide a full spectrum of logistics support and sustainment. Their designs are standardized and will usually consist of five to eight companies, which can be task-organized to support TSC opening, distribution, area sustainment, or life-support functions, to include SOF sustainment needs.

1-22. SOF missions must remain a prime consideration in the functions of ARSOF and theater of operations logistics units. Logistic resources and priorities must be tailored to the changing ARSOF environment and to support unified land operations. Logistics units must be flexible and responsive enough to operate from any support-base arrangement. They must be able to operate and survive in hostile environments and accomplish their missions.

PRINCIPLES OF SUSTAINMENT

1-23. The principles of sustainment are critical to guiding the success of generating combat power, strategic and operational reach, and endurance. These principles are anticipation, responsiveness, simplicity, economy, survivability, continuity, improvisation, and integration. Sustaining SO missions throughout any operation or event is important to success. Tailored SOF packages maximize the capability of initial-entry forces consistent with the mission and the requirement to project, employ, and sustain the force. ARSOF sustainment planners must work hand-in-hand with SOF operational planners to synchronize sustainment to enable operational reach. Endurance is the ability to employ combat power anywhere for protracted periods. Endurance stems from the ability to generate, protect, and sustain a force, regardless of how far away it is deployed, how austere the environment, or how long land power is required. Providing sustainment to support operations consistent with the commanders' intent and requirements is critical to ARSOF projection and success.

ANTICIPATION

1-24. Anticipation is being able to foresee future operations and events, and identify and maintain the right support to sustain the force, whether they are contingency operations or theater security cooperation plan (TSCP) events. Sustainment planners must anticipate future events and sustainment requirements, and understand the commander's intent to best ensure uninterrupted support to the force. Accurate forecasts of operations are needed to develop a force that is strategically responsive, deployable, and fully capable of performing missions it is likely to receive. Anticipation enhances endurance. Anticipation involves making the most effective and efficient use of available resources. No planner can fully predict events of the future. Anticipating sustainment requirements means staying abreast of current operation plans (OPLANs), continuously assessing requirements, and tailoring ARSOF and Army sustainment to meet an ever-changing environment.

RESPONSIVENESS

1-25. Responsiveness is the ability to meet changing requirements on short notice. It is providing the right support in the right place at the right time. It includes the ability to see operational requirements. Employing appropriate information systems enables commanders to make rapid decisions. Responsiveness involves identifying, accumulating, and maintaining the minimum assets, capabilities, and information necessary to meet rapidly changing requirements. A responsive sustainment system is crucial to maintaining endurance; it provides the ARSOF commander with flexibility and freedom of action. Through responsive sustainment, commanders maintain operational focus and pressure, set the tempo of friendly operations to prevent exhaustion, replace ineffective units, and extend operational reach. Responsiveness rests on anticipation.

SIMPLICITY

1-26. Simplicity is defined as a minimum of complexity in logistics operations. Complexity introduces confusion into an already chaotic environment. Simplicity fosters efficiency in planning and execution, and allows for more effective control over logistics operations. Clarity of tasks, standardized and interoperable procedures, and clearly defined command relationships contribute to simplicity. Simplicity enables economy and efficiency of sustainment resources, ensuring effective sustainment operations. Because of the size and nature of most SOF missions, simplicity is one of the key principles to sustainment mission success.

ECONOMY

1-27. Economy means providing effective sustainment using the fewest resources within acceptable levels of risk. Resources are always limited. The commander achieves economy by prioritizing and allocating resources. Economy reflects the reality of resource shortfalls, while recognizing the inevitable friction and uncertainty of military operations. The modular force was designed to achieve economies of scale. Joint interdependence is one method used to achieve this goal. Reliance on host-nation support (HNS) and operational contracting are other methods used that contribute to economy. Effective use of information technology allows commanders to anticipate requirements, track resources, and make decisions enabling economy of resources.

SURVIVABILITY

1-28. Survivability is the ability to protect sustainment functions from destruction or degradation. Survivability is a function of protection. It consists of those actions to prevent or mitigate hostile actions against personnel, resources, facilities, and critical information. Integrating protection with operational plans is critical to sustainment survivability. Economy contributes to survivability by minimizing the sustainment resources that require protection. Dispersion and decentralization of sustainment operations may also enhance survivability. The ARSOF commander may have to balance risk with survivability in considering redundant capabilities and alternative support plans. Survivability of sustainment ensures operational reach and endurance.

CONTINUITY

1-29. Continuity is the ability to maintain uninterrupted support during all phases of campaigns and operations. Continuity is essential to strategic and operational reach and endurance. The generating force must maintain a strong link to operating forces to ensure continuity in the flow of sustainment. ARSOF theater of operations sustainment planners must work hand-in-hand with operation planners to understand and synchronize requirements over the entire course of the operation. A disruption in the flow of sustainment could result in a loss of reach, thus causing a pause or early culmination of an operation and thereby loss of the initiative.

IMPROVISATION

1-30. Improvisation is the ability to adapt sustainment to changing situations and missions. It includes creating, inventing, arranging, or fabricating what is needed from what is on hand. Today's high-tech operational environment involves an enemy that quickly evolves and adapts to changing scenarios and environments. More than ever, this type of environment requires sustainment commanders, their staffs, and Soldiers to quickly adjust and use any means possible to maintain a continuous operation. Sustainment commanders must visualize complex operations and understand what is possible at the tactical level. These skills enable commanders to improvise operational and tactical sustainment actions when enemy actions attempt to disrupt sustainment operations.

INTEGRATION

1-31. Integration is the most critical principle. It is the deliberate coordination and synchronization of sustainment within any operation and at each level of war. ARSOF integrate their sustainment operations with other components of the joint force to benefit from each Service component's competencies and resources. Integration requires a thorough understanding of the commander's intent and synchronization of sustainment with the concept of operations. Integration of sustainment with joint forces (joint interdependence) allows efficiencies through economies of scale. It ensures the highest priorities of the joint force are met first and avoids redundancy. It also eliminates wasteful competition for scarce strategic-lift and theater of operations resources.

SUSTAINMENT WARFIGHTING FUNCTION

1-32. As previously discussed, sustainment is the comprehensive term covering the functions of logistics, personnel services, and health service support (HSS). The sustainment warfighting function is one of six Army warfighting functions (movement and maneuver, fires, protection, sustainment, mission command, and intelligence) that produce combat power. The sustainment warfighting function is defined as the related tasks and systems that provide support and services to ensure freedom of action, extend operational reach, and prolong endurance (ADP 3-0, *Unified Land Operations*). Warfighting functions make up elements of combat power and are tied together by information and leadership. A brief description of the sustainment warfighting function follows.

LOGISTICS

1-33. Logistics is planning and executing the movement and support of forces. It includes those aspects of military operations that deal with design and development, acquisition, storage, movement, distribution, maintenance, evacuation, and disposition of materiel; movement, evacuation, and hospitalization of personnel; acquisition or construction, maintenance, operation, and disposition of facilities; and acquisition or furnishing of services (Joint Publication [JP] 4-0, *Joint Logistics*). (ADRP 4-0 provides a full discussion on logistics operations.)

Supply

1-34. Supply is the procurement, distribution, maintenance while in storage, and salvage of supplies, including the determination of kind and quantity of supplies. Supply consists of a producer phase and a consumer phase (JP 4-0). (ADRP 4-0 provides a full discussion on supply functions.)

Field Services

1-35. Field services are essential services for enhancing the quality of life of Soldiers. They include clothing repair and exchange, laundry and shower support, mortuary affairs, aerial delivery, food services, billeting, and sanitation. All field services receive the same basic Army-wide priority, but the commander decides which are most important. (ADRP 4-0 provides a full discussion on field services.)

Maintenance

1-36. Maintenance is defined as all actions taken to retain materiel in a serviceable condition or to restore it to serviceability. Army maintenance consists of two levels of maintenance: field and sustainment maintenance. It includes inspection, testing, servicing, and classification as to serviceability, repair, rebuilding, recapitalization, reset, and reclamation. It also includes all supply and repair actions taken to keep a force in condition to carry out its mission. From a proponent perspective, ammunition is viewed as an ordnance function along with maintenance. From a warfighting function perspective, ammunition will be discussed under sustainment. (Army Tactics, Techniques, and Procedures [ATTP] 4-33, *Maintenance Operations*, provides a full discussion on maintenance operations.)

Transportation

1-37. Transportation is the moving and transferring of units, personnel, equipment, and supplies to support the concept of operations. Transportation plays a key role in facilitating force projection and sustainment. Transportation incorporates military, commercial, and multinational capabilities. Transportation assets include motor, rail, air, and water modes and units; terminal units, activities, and infrastructure; and movement-control units and activities. (ATP 4-16, *Movement Control*, provides a full discussion on transportation operations.)

General Engineering

1-38. General engineering are those engineering capabilities and activities, other than combat engineering, that modify, maintain, or protect the physical environment. Examples include the construction, repair, maintenance, and operation of infrastructure, facilities, lines of communication and bases; terrain modification and repair; and selected explosive hazard activities. Engineering provides construction support, real estate planning and acquisition, and real property maintenance responsive to environmental considerations. (FM 3-34.400, *General Engineering*, provides a full discussion on general engineering.)

Operational Contract Support

1-39. Operational contract support is the process of planning for and obtaining supplies, services, and construction from commercial sources in support of operations along with the associated contractor management functions. Deployed U.S. forces rely increasingly on contracting to supplement organic sustainment capabilities and on contractors to perform a growing percentage of many sustainment functions. (JP 4-10, *Operational Contract Support*, and ATTP 4-10, *Operational Contract Support Tactics, Techniques, and Procedures*, provide a full discussion on operational contract support.)

PERSONNEL SERVICES

1-40. Personnel services are those sustainment functions related to Soldier welfare, readiness, and quality of life. Personnel services include human resources support, financial management, legal support, religious support, and band support. Personnel services complement sustainment by providing the personnel required so that the unit may be best prepared to accomplish its assigned mission. Personnel services complement logistics by planning for and coordinating efforts that provide and sustain personnel.

Human Resources Support

1-41. Human resources support includes the human resources functions of manning the force, human resources services, personnel support, and human resources planning and operations. Human resources support maximizes operational effectiveness and facilitates support to Soldiers, their families, Department of Defense (DOD) civilians, and contractors who deploy with the force. Human resources support includes personnel readiness management; personnel accountability; strength reporting; personnel information management; casualty operations; essential personnel services, band support, and postal operations; reception, replacement, return-to-duty, rest and recuperation, and redeployment operations; morale, welfare, and recreation; and human resources planning and staff operations. (FM 1-0, *Human Resources Support*, provides a full discussion on human resources support.)

Financial Management

1-42. Financial management is comprised of two mutually supporting core functions: finance operations and resource management (RM). Finance operations include developing policy, providing guidance and financial advice to commanders, disbursing support to the procurement process, banking and currency, accounting, and limited pay support. RM operations include providing advice to commanders; maintaining accounting records; establishing a management internal control process; developing resource requirements; identifying, acquiring, distributing, and controlling funds; and tracking, analyzing, and reporting budget execution. (FM 1-06, *Financial Management Operations*, provides a full discussion on financial management.)

Legal Support

1-43. Legal support is the provision of professional legal services at all echelons. Legal support encompasses all legal services provided by judge advocates and other legal personnel in support of units, commanders, and Soldiers in an AO and throughout unified land operations. Judge Advocate General's Corps personnel assist Soldiers in personal legal matters and advise commanders on a wide variety of operational legal issues. These include the law of war, rules of engagement, lethal and nonlethal targeting, treatment of detainees and noncombatants, fiscal law, claims, contingency contracting, the conduct of investigations, and military justice. (FM 1-04, *Legal Support to the Operational Army*, provides a full discussion on legal support.)

Religious Support

1-44. Religious support facilitates the free exercise of religion, provides religious activities, and advises commands on matters of morals and morale. The First Amendment of the U.S. Constitution and Army Regulation (AR) 165-1, *Army Chaplain Corps Activities*, guarantee every American the right to the free exercise of religion. Commanders are responsible for those religious freedoms within their command. Chaplains perform and provide religious support in the Army to ensure the free exercise of religion. (FM 1-05, *Religious Support*, provides a full discussion on religious support.)

Band Support

1-45. Army bands provide critical support to the force by tailoring music support throughout military operations. Music instills in Soldiers the will to fight and win, foster the support of U.S. citizens, and promote national interests at home and abroad. (ATTP 1-19, *U.S. Army Bands*, provides a full discussion on band support.)

HEALTH SERVICE SUPPORT

1-46. *Health service support* is all services performed, provided, or arranged to promote, improve, conserve, or restore the mental or physical well-being of personnel, which include, but are not limited to, the management of health services resources, such as manpower, monies, and facilities; preventive and curative health measures; evacuation of the wounded, injured, or sick; selection of the medically fit and disposition of the medically unfit; blood management; medical supply, equipment, and maintenance thereof; combat and operational stress control; and medical, dental, veterinary, laboratory, optometric, nutrition therapy, and medical intelligence services (JP 4-02, *Health Service Support*).

EXPEDITIONARY LOGISTICS IMPERATIVES

1-47. The following paragraphs discuss the TSOCs and/or SOTF expeditionary logistics imperatives. Although the imperatives may not apply to all types of SOF requirements, ARSOF commanders must include the applicable imperatives in their mission planning and execution, especially when developing the concept of support.

UNDERSTANDING THE OPERATIONAL ENVIRONMENT

1-48. ARSOF cannot dominate the operational environment without first gaining a clear understanding of theater of operations dynamics, theater of operations infrastructure, and sustainment capability. ARSOF logistics systems, processes, and organizations must be agile, versatile, flexible, and responsive. Being agile is being globally responsive and rapidly deployable to any environment. Versatile means undertaking a variety of missions across the range of military operations. Flexible means being flexible enough to rapidly shift focus from one mission to an entirely different one without retraining, refitting, or reorganizing. Responsive means being forward-deployed or being a rapidly deployable force that can respond swiftly with sustainable combat power. Expeditionary ARSOF must also be force-entry-capable and possess "come as you are" capability. Finally, they must be sustainable, allowing for continuous operations in an austere environment without HNS and without reliance upon preexisting infrastructure.

UNITY OF EFFORT

1-49. Unity of effort is the coordination application of all logistics capabilities focused on the TSOC's and/or SOTF commander's intent, and is the most critical of all logistics outcomes. Achieving unity of effort requires the optimal integration of SOF, joint, multinational, interagency, and nongovernmental logistics capabilities, built around three enablers:
- Appropriate organizational capabilities and authorities provide the means to effectively and efficiently execute SOF logistics.
- Shared awareness across the logistics domain drives unity by focusing capabilities against the SOF commander's most important requirements. The effective integration of SOF priorities and

Chapter 1

the continuous optimization of those priorities in space and time are key tasks requiring shared awareness.
- Common measures of performance drive optimization across processes supporting the SOTF. Clearly defined SOF logistics processes, well-understood roles and accountabilities of the players in the processes, and shared SOF metrics frame this enabler.

RAPID AND PRECISE RESPONSE

1-50. Rapid and precise response is defined by the ability of the supply chain to effectively meet the constantly changing needs of the task force. The lack of key supplies, regardless of the reason, acts to undermine readiness and increase mission risk. The following performance measures indicate how well the supply chain is responding to identified requirements for the SOTF:
- Speed and accuracy are core aspects for responsiveness to the TSOCs and the SOTF.
- Reliability is the ability of the supply chain to provide predictability or time-definite delivery.
- Visibility provides rapid and easy access to order information.
- Efficiency is directly related to the supply chain's footprint.

DOMAIN-WIDE VISIBILITY

1-51. Domain-wide visibility is the ability to see the requirements, resources, and capabilities across the logistics domain, both SOF and Army. Three fundamental enablers frame the ability to achieve this imperative:
- Connectivity, which is 24-7 network access, reaching globally—back, forward, and laterally throughout the network—to synchronize and coordinate efforts of supporting SOF, to include interagency participants, multinational partners, and host nations (HNs).
- Standard enterprise data architecture is the foundation for effective and rapid data transfer, and forms the fundamental block to enable a common logistical picture.
- A global and combatant command focus over the processes that deliver support to the TSOCs and/or SOTF is paramount to optimizing Service component and SOF logistics. Logistics support to any SOF element is a global business, and an operational perspective that operates below this level will deliver less-than-acceptable readiness.

Chapter 2
Army Special Operations Forces Logistics Support Framework

USASOC has aligned its ARSOF logistics organizations and activities in support of the U.S. Army's concept of modularity and force projection. This alignment allows ARSOF to integrate organic support elements within the theater of operations support structure for responsive and continuous sustainment of ARSOF units. Therefore, the planning and execution of logistics support must be well-nested within the CCDR's concepts of operation and tailored to interface with the theater of operations logistics structure.

ARMY SUPPORT STRUCTURE

2-1. ARSOF missions in support of USSOCOM are inherently joint. Although Title 10 does require each Service to provide its own logistics support, authority is available through other means to conduct joint sustainment. The CCDR exercises control over subordinate commands through directive authority for logistics (DAFL), and ARSOF may receive or provide logistics support from or to other Services. ARSOF typically operate in a joint environment, and CUL support may be controlled and provided by other means. Other sources and authority for CUL support, aside from the ASCC Title 10, include—

- DOD executive agent directives and instructions.
- Interservice support agreements (ISSAs).
- Acquisition and cross-servicing agreements (ACSAs).
- CCDR's OPLANs, operation orders (OPORDs), and directives.

2-2. Options for executing logistics support to a joint force include any combination of the following:

- Single Service component dedicated support—each Service component supports its own forces.
- Lead Service or agency support—a lead Service or agency provides common user/item support to one or more Service components and governmental or other organizations.

2-3. Strategic support of sustainment is available through the Defense Logistics Agency, Defense Finance and Accounting Service, U.S. Army Human Resources Command, and the U.S. Army Finance Command.

2-4. The generating force's support consists primarily of the United States Army Training and Doctrine Command, United States Army Materiel Command (USAMC), United States Army Forces Command, and the United States Army Medical Command. Elements of the strategic base, such as the USAMC logistics support element, deploy to the JOA and are integrated into the overall logistics structure to provide support at the operational level and, when required, at the tactical level.

2-5. The operational Army support structure is made up of modular forces and units such as the ASCC, TSC, ESC, Army sustainment brigade, Army field support brigade, and brigade support battalion. USASOC serves as the ASCC for ARSOF but functions similarly to the Army's generating force and serves as a global resourcing command in support of USSOCOM, a combatant command. The theater of operations ASCCs provide ARSOF logistics support within a CCDR's AOR. The GSBs operate at the tactical level and provide support that is equitable to brigade support battalions but are not as robust in capability.

LOGISTICS INTEGRATION INTO OPERATIONS

2-6. The operations process of planning, preparation, execution, and assessment applies to ARSOF sustainment supporting unified land operations. Similar to Army conventional forces, ARSOF sustainment commanders and staff must synchronize and integrate the sustainment plan with the ARSOF mission plan, while simultaneously integrating and synchronizing with the theater of operations logistic concept of support. However, unlike Army conventional forces, ARSOF will more heavily rely on the theater of operations logistic capabilities. There are several contributing factors that create a greater need for direct, reinforcing, and backup logistics support. Because of the nature of ARSOF employment concepts, ARSOF operate for extremely short or long durations, in small team elements, dispersed across the CCDR's AOR along extended lines of operations, and in typically austere environments or immature theater of operations logistic structures.

2-7. To effectively execute the plan, ARSOF sustainment commanders and staff must take procedures to prepare for the execution of the operation. One of the means for preparation is sustainment preparation of the operational environment. This may entail HN agreements, ACSAs, and contracting. Other preparations include pre-positioned stocks, predeployment site surveys, facility and port assessments, medical preparations, and an array of rehearsals. The execution of sustainment includes the deployment and distribution processes.

2-8. In executing logistic operations, ARSOF logisticians will optimize operational reach and endurance throughout the five processes of force projection—mobilization, deployment, employment, maintaining personnel and materiel, and redeployment. Distribution is the largest single process in the execution phase, controlled through distribution management centers and the use of in-transit visibility enablers. ARSOF logisticians at all levels of operations must synchronize and integrate with the theater of operations distribution network in order to achieve maximum effectiveness of both organic and theater of operations support distribution assets. Commanders must anticipate the possible need for reconstitution as part of execution operations and plan accordingly. The continual assessment of sustainment operations ensures mission success and allows ARSOF sustainment commanders and staff to adjust to changing situations, as required, when the theater of operations logistic structure matures or decreases.

SUPPORT RELATIONSHIPS

2-9. ARSOF operating and logistical structures differ vastly from Army conventional forces. The SFGs are the only units that have any type of organic direct support capability as discussed in Chapter 1. The GSB within the SFG provides direct support to the SFG or to the JSOTF elements when directed by the TSOC. The Ranger Regiment, ARSOAC, CA brigade, and MISOC do not possess any organic direct support assets. The ARSOAC, typically task-organized under a joint special operations air component, will be provided direct support by the joint special operations air component's direct support elements and the CUL-designated provider. The CA brigade and MISOC will be supported through their task organization's direct support elements; for example, the GSB when the CA or MISOC element is task-organized under a JSOTF. For all ARSOF units, the direct support arrangement is METT-TC driven, and direct support may be provided on an area basis by the Army sustainment brigade's CSSB. General support to ARSOF units will be provided by the ASCC. Figure 2-1, page 2-3, is an example of a typical ARSOF sustainment structure.

2-10. The 528th SB(SO)(A)'s mission and structure is significantly different from a conventional force Army sustainment brigade and primarily supports ARSOF globally, through planning, synchronizing, and integration of operational logistics. The ALEs and ASPO cells conduct planning, synchronization, and integration of operational logistics with the TSOCs, JSOTFs, ASCCs, GSBs, RSOD, and the CA and MISOC S-4s. However, the 528th SB(SO)(A) can deploy a tailored brigade HQ for mission command of attached CSSBs in support of ARSOF for a limited duration, serving as interim mission command until theater of operations logistics structure develops.

Army Special Operations Forces
Logistics Support Framework

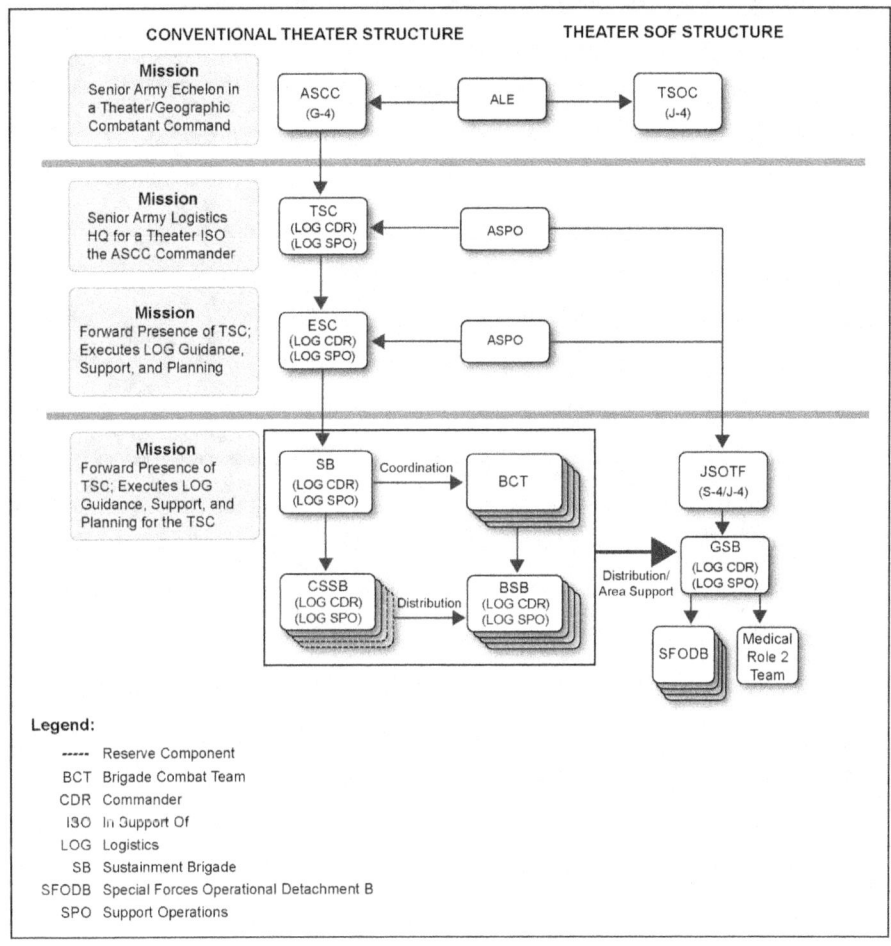

Figure 2-1. Typical ARSOF sustainment structure

LOGISTICS PLANNING AND PREPARATION

2-11. The principal focus of logistics planning is at the operational level. The challenge for logisticians is to link strategic resources to tactical unit requirements. The objective of logistics planning is to fully integrate and coordinate support and operational execution to ensure sustained operational readiness. Planning logistics support links the mission, commander's intent, and operational objectives to core logistic capabilities, procedures, and organizations. Logistics planning defines processes and procedures to establish an effective concept for logistics support.

Chapter 2

KEY PLANNING ROLES, RESPONSIBILITIES, AND RELATIONSHIPS

2-12. The USASOC DCS G-4 maintains a strategic-to-operational-level planning focus and performs contingency, crisis, mobilization, exercise, and reconstitution planning in coordination with USSOCOM, CCDRs supporting subcombatant commands (theater of operations ASCCs), Department of the Army, USAMC, and TSOCs. Theater-specific support planning functions include limited sustainment planning and theater of operations transportation and ammunition requirements. Within USASOC, the G-4 conducts staff planning and coordination, assisting the G-3 and G-8 to validate and source capabilities requirements. Concurrently, the G-4 conducts planning efforts with the 528th SB(SO)(A), U.S. Army Special Forces Command G-4, and TSOC J-4s.

2-13. The U.S. Army Special Forces Command Assistant Chief of Staff, G-4, develops operational-level plans and orders in support of the SFGs within USASOC for ARSOF mobilization, predeployment, deployment, and redeployment operations. The main planning focus for theater of operations support requirements includes ammunition and transportation requirements. During the planning and preparation phase, the U.S. Army Special Forces Command G-4 captures logistic shortfalls and requirements, cross-leveling within the command when feasible, and ensures external support requirements are passed to the USASOC and the 528th SB(SO)(A) for resolution or sourcing.

2-14. The 528th SB(SO)(A) ALE established within each CCDR's AOR is a key strategic- and operational-level logistics planner for ARSOF missions. Within the adaptive planning and execution (APEX) joint logistics planning process, at the combatant command level, planning begins with the receipt of strategic guidance and continues as the CCDR conducts mission analysis. During this step, the CCDR J-4 staff and Service-component logisticians begin to develop a theater of operations logistics overview, which becomes the concept of logistics framework.

2-15. The ALE, in concert with the 528th SB(SO)(A)'s Plans Section, TSOC J-4, and ASCC G-4, assists with establishing and updating a standing logistics estimate of each theater of operations derived from the theater of operations logistics overview. The ALE remains integrated in the CCDR's, TSOC's, and ASCC's logistics planning processes during plan development and assessment. The 528th SB(SO)(A) Operations Division, Plans Section, establishes logistics and FHP planning conferences during the planning and preparation phase to develop the operational-level ARSOF concept of support in concert with deploying ARSOF sustainment planners, to include the GSB, RSOD, and brigade S-4s.

2-16. The GSB commander and ASPO cells, brigade S-4s, and RSOD within the task organization of the ARSOF or JSOTF conduct the military decisionmaking process and parallel logistics planning. They also identify logistics requirements and shortfalls based on their organic logistics capabilities and command and support relationships.

KEY SPECIAL OPERATIONS PLANNING CONSIDERATIONS DURING THE MILITARY DECISIONMAKING PROCESS

2-17. Because of the nature of ARSOF missions, logistics planning considerations differ somewhat from Army conventional forces logistics planning and some of the key planning factors may be more significant because of their impact on missions. ARSOF logistics units can support a variety of humanitarian, civil, and security assistance programs. Deployed ARSOF units are usually in isolated and austere locations. Distribution and resupply are key considerations that may require ARSOF-unique capabilities. Sustainment planning and preparation efforts must first consider the existing infrastructure in the theater of operations. Using this infrastructure as a baseline, planners then integrate, consolidate, and cross-level resources to maximize logistics support. Planning considerations include but are not limited to the following:

- Identify SOF-peculiar logistics support plans for SO-specific equipment.
- Determine if austere-based operations are likely in the early stages of deployments; ARSOF will normally be required to establish separate intermediate staging bases (ISBs) and eventually expand the number of support bases to meet mission requirements.
- Minimize the logistic footprint.
- Maximize the use of existing fixed facilities.

- Limit logistics requirements to mission essentials and acceptable risk.
- Rely on air lines of communications (ALOCs) for rapid resupply.
- Concentrate maintenance on returning major end items to service.
- Anticipate high attrition of supplies while performing missions in denied areas.
- Identify to the ASCC as early as possible those items that require operational floats or other special logistics arrangements.
- Make maximum use of HNS, including local and third-country resources, to include facilities, lines of communications, ports, airfields, communication systems, and services.
- Conduct threat and risk assessment.
- Determine the impact of topography, climate, and external factors on the logistical system, to include available lines of communications, including waterways, roads, railroads, pipelines, and airways.
- Purchase or contract local supplies, facilities, utilities, services, and transportation support systems.
- Develop or improve the HN logistics capabilities intended for eventual transfer of responsibility to the supported nation.
- Develop intertheater and intratheater airlift and sealift to deliver supplies.
- Determine legal considerations for providing support, materiel aid, and health services to indigenous personnel and allied or HN military forces and civilians.
- Determine relationship with units in the Army Force Generation.

Note: Appendix A provides additional logistical planning considerations.

ARMY FORCE GENERATION

2-18. Army Force Generation is the Army's system for generating land power capabilities that respond to the operational needs of joint force commanders and sustaining those capabilities as long as required. It is a shift from tiered readiness to cyclic readiness and represents a change in the way the Army generating force performs its Title 10 functions. The generating force resets operational forces upon redeployment. Units are in a state of Train, Ready, or Reset. ARSOF need support from conventional forces in areas such as forward surgical teams, explosive ordnance disposal, and CSSBs and rely on the Army Force Generation process to align conventional forces to ARSOF units.

SPECIAL OPERATIONS FORCE GENERATION

2-19. Beginning in October 2011, Special Operations Force Generation (SOFORGEN) became the key process within USASOC that synchronizes component subordinate commands and component subordinate units and the staff to provide sustainable and predictable forces with subregional expertise to the GCCs while protecting and preserving the critical characteristics of ARSOF. The process is a paradigm shift in force management from "demand-based" to "supply-based." SOFORGEN also identifies demand signals for conventional forces to be aligned in the Army Force Generation process.

SOURCES OF SUPPLY AND SUPPORT

2-20. ARSOF are supplied from a variety of sources. The relative reliability of each source is theater of operations-dependent. ARSOF commanders must anticipate their logistical requirements. They must coordinate with their supporting ASCC to determine which source of sustainment can best meet their needs. Among these sources of supply are Army pre-positioned stocks/operational project stocks, joint operational stocks (JOS), war reserve sustainment stocks, HNS, and contracting.

Army Pre-positioned Stocks

2-21. Select ARSOF units have Department of the Army authorization to receive Army pre-positioned stocks equipment when they deploy from the continental United States (CONUS) to their theaters of operation. Deploying units must determine existing Army pre-positioned stock shortages before deploying and must deploy with those items, as well as items not authorized for pre-positioning. Units should update their deployment plans upon receipt of their annual Army pre-positioned stocks authorization document in coordination with the USASOC G-4, Plans and Analysis Division.

Operational Project Stocks

2-22. ARSOF units use operational project stocks to obtain required supplies and equipment above their normal allowances to support contingency operations and war plans. The theater of operations CCDR or TSOC may also establish and maintain operational project stocks to support joint SO. Operational project stocks are restricted to the minimum-essential types and quantities of supplies, equipment required for successful execution of the total plan, or prescribed portions of the plan. Stocks normally include only standard items listed on the war reserve stockage list. ARSOF commanders must justify the inclusion of nonstandard items in project stocks.

Joint Operational Stocks

2-23. The JOS is a supply of centrally managed and maintained USSOCOM mission-essential equipment. JOS are available on a loan basis to USSOCOM units. The JOS Catalog consists of weapons, optics and night-vision equipment, communications, force protection, bare-base, and limited specialty equipment.

2-24. JOS consist of four basic categories:
- *Mission-essential and personal protection.* This category includes equipment routinely issued for mission and training support, to include optical and night-vision devices, weapons, radios, and personal protection devices.
- *Bare-based equipment.* This category includes equipment necessary to augment or supplement unit assets for training and mission support, to include tents, environmental control units, and light sets.
- *Contingency operations.* This category includes equipment held for specific or unique mission requirements. This category is not applicable for training issue.
- *Humanitarian assistance support.* This category includes limited support to a survey team or a small deployed element, to include Class X items.

2-25. Units submit their JOS requests through command channels to USSOCOM HQ, where requests are validated and then sent to the special operations forces support activity (SOFSA), which processes the approved requests. Personnel then package and ship the equipment directly to the requesting unit. The JOS Catalog describes procedures for obtaining equipment, the basic loan request process, and services available. The Web site can be accessed at https://ssavie.cocom.mil. Appendix B provides additional information on JOS.

OTHER SOURCES OF SUPPLY AND SUPPORT

2-26. In addition to the JOS, there is the USASOC Redistribution Center, which is a contractor-operated storage facility located in Lexington, Kentucky. The USASOC Redistribution Center stores, refurbishes, and reissues acquired equipment inclusive of bare-base assets; weapons system replacement operations/operational readiness float assets; unit administrative storage items, to include statement of requirement (SOR) items, ground mobility vehicle rebuilt parts and materiel, and medical supplies and equipment; and the command's weapons system replacement operations/operational readiness float assets in support of contingency operations. The USASOC DCS, G-4, has primary staff responsibility for the USASOC Redistribution Center.

2-27. The operational needs statement (ONS) is a request document for materiel to correct a deficiency, improve a capability, or request HQDA to procure a new or emerging capability that enhances mission accomplishment. An ONS is a request for HQDA validation, authorization, and sourcing of a perceived requirement. The other type of request is the equipment sourcing document (ESD), which is a request for the sourcing of a requirement that has already been validated with a modified table of organization and equipment (MTOE) authorization by HQDA. Numerous methods and formats for requesting validated equipment exist, including the digital process within the ECOP database. The Army ECOP is a SECRET Internet Protocol Router Network (SIPRNET)-based start-to-finish database for submitting requirements and tracking them through the validation and solution approval and sourcing process. The ONS and ESD are submitted to HQDA via the Army ECOP. ARSOF can submit an ESD and/or an ONS prior to deployment or when deployed. The ESD/ONS can be submitted at the component subordinate unit level (O-6 unit-level endorsement) and requires endorsement and approval by the supporting ASCC. Deployed ARSOF should submit requirements through the TSOC for endorsement prior to staffing with the theater of operations ASCC. Online ECOP database staff coordination should take place at all levels; requesting units can defend their requirements.

2-28. A deployment statement of requirement (DSOR) identifies those material and nonmaterial requirements the unit cannot satisfy with its organic assets or capabilities needed to support a mission or task. The unit staffs the DSOR and submits it to the component subordinate unit's commander for signature, who then submits it to the G-3/S-3 for validation. Funding for the DSOR may come from programmed dollars and/or be submitted as an unfinanced requirement. In general, the command's fiscal year general and specific guidance will dictate how DSORs will be funded. The Combat-Mission Needs Statement (C-MNS) is the rapid validation, approval, and fielding of critical (mission failure or loss of life), new, or existing SOF materiel capability. The C-MNS is used for SO-peculiar materiel whereas the ONS or ESD is used for an Army-common solution. The C-MNS is not an unfinanced requirement or supplemental process, nor is it a means to circumvent or accelerate the normal Joint Capabilities Integration and Development System (JCIDS) or other USSOCOM-established materiel solution process. The C-MNS process is very similar to the ONS process in that it begins with a need, then a recommended solution (mission analysis), and eventually the sourcing of that solution. The C-MNSs are generated by the SOF unit to the TSOC, which then sends the requirement to the SOF component to see if it can satisfy the SO-peculiar requirement. If the SOF component cannot satisfy the SO-peculiar requirement, then it endorses the request and sends it on to USSOCOM for further staffing and, if approved, to the eventual resourcing activity.

2-29. HNS is civil or military assistance rendered by a nation to foreign forces within its territory in support of the full range of military operations through agreements mutually concluded between nations. HNS includes all civil and military support a nation provides to multinational forces located in its sovereign territory. HNS is not the same as contractor support. Multinational forces and contractors may have an impact on the ability of a HN to provide HNS. This situation must be deconflicted during the logistics and personnel support planning of the operation, especially in areas of limited resources or where operational security is of concern. HNS is the preferred means to meet unresourced support requirements within acceptable risk limits. Quality of HNS is theater of operations- and situational-dependent. Availability of support depends upon the geographical area, prior agreements with nations in the region, and the nation's ability or willingness to provide support. In some theaters of operation, agreements may exist between the United States and the HN. Potential HNS agreements should address labor support arrangements for port and terminal operations and the use of available transportation assets in-country. The agreements should also address the use of bulk petroleum distribution and storage facilities, the possible supply of Class III (Bulk) and Class IV, and the development and use of field services.

2-30. Contractor support can mitigate shortfalls of logistics support requirements. A contracting support plan must identify operational-specific contractor integration policies, procedures, and requirements so the Service components, joint operational contracting support command (if established), SOF, and Defense Logistics Agency can integrate applicable contracting plans. The CCDR's contracting support plan will identify organizational options for use during theater of operations joint operations (for example, Service support to own forces, a lead Service for logistics contracting, or a joint contracting organization) across the full range of military operations or phases of operations. There are three broad types of contracted support: theater support, external support, and systems support. For ARSOF sustainment operations planning and execution, the theater of operations and external contracting support must be properly incorporated and synchronized with the overall logistics plan. The operational-level planning, integration, and synchronization conducted by the 528th SB(SO)(A)'s ALEs and ASPO cells with the TSC support operations (SPO), TSC/ESC SPO, the Army forces G-4, and the supporting Army field support brigade's contingency contracting officer (CCO) team and/or the logistics civilian augmentation program support unit is key in leveraging the ASCC's general support in the theater of operations. SO-peculiar contracting support is provided by the USSOCOM Contracting Activity. Chapter 11 includes more detailed discussion on contracting support.

GEOGRAPHIC COMBATANT COMMANDERS' THEATER LOGISTICS ENVIRONMENT

2-31. The regional environments and logistical staffing processes and structures are unique in each GCC's AOR and each has a TSOC. The TSOC is a subcombatant command that exercises operational control of theater of operations Army, Navy, Air Force, and Marine SOF. The GCC also has a number of other commands and organizations promoting security and peaceful development in their regions, such as the other Service components, both SOF and Army conventional forces, regional military partners, and U.S. Government agencies, including the Department of State's (DOS's) United States Agency for International Development (USAID) (providing economic and humanitarian assistance). The GCC may also have a JTF and/or JSOTF under his command for contingency operations or in support of a particular concept plan (CONPLAN) or OPLAN. The following paragraphs provide a brief description of responsibilities, organizations, and processes for the GCC's AORs.

UNITED STATES PACIFIC COMMAND

2-32. In the United States Pacific Command (USPACOM) AOR, planning for SOF contingency operations or TSCP events is a challenge for any ARSOF logistics planner. In the Asia Pacific region, there are 16 time zones comprising 50 percent of the earth's surface and 60 percent of the world's population. The USPACOM region is home to the world's six largest armed forces and five of the seven worldwide U.S. mutual defense treaties. This region also has major social and economic implications. Logistics planning can take on a wide spectrum of support operations, from mature theater of operations like Korea to the isolated islands of the Philippines.

2-33. The USPACOM GCC's lead subcombatant command for SOF missions is the Special Operations Component, United States Pacific Command, better known as SOCPAC. The ASCC for the USPACOM GCC is the United States Army, Pacific Command (USARPAC), with Title 10 responsibility as the lead Service providing common user/items support to Army forces deployed in the USPACOM AO. USARPAC's senior Army logistics command, the 8th TSC and its subordinate elements, 19th ESC, and other units are responsible for executing theater of operations opening, distribution, and sustainment, to include sustainment of deployed SOF. The USPACOM GCC also has a number of other commands and organizations promoting security and peaceful development in the Asia Pacific region, such as the Joint Interagency Task Force-West (counterdrug).

2-34. Contractor support plays a vital role in sustaining contingency requirements and is currently an effective force multiplier for the current standing JSOTF in the region. The Navy is USPACOM's executive agent for logistics as designated by DOD and is providing contractual support to the region's standing JSOTF. Any one of the Service components and activities with contracting authority can plan and award Service support contracts in support of contingency operations and promoting security or peaceful development.

UNITED STATES SOUTHERN COMMAND

2-35. The United States Southern Command (USSOUTHCOM) GCC's lead subcombatant command for SOF missions is the Special Operations Command, South (SOCSO), better known as SOCSOUTH. USSOUTHCOM conducts numerous planning conferences with its supporting elements, to include the SOF components. The theater of operations ASCC for the USSOUTHCOM GCC is the United States

Army, Southern Command (USARSO), with Title 10 responsibility as the lead Service providing common user/items support to Army forces deployed in the USSOUTHCOM AO. USARSO's senior Army logistics command, the 377th TSC (USAR), and its sustainment force pool (ESCs) consisting of several USAR units, are responsible for executing theater of operations opening, distribution, and sustainment, to include sustainment of deployed SOF. The USAR theater of operations sustainment support structure is a challenge, especially when dealing with crisis action planning and sustained logistics support for continuous operations. The challenge lies with USAR mobilization timelines and continuity of effort in a theater of operations that sees its share of operations and security cooperation events. The USSOUTHCOM GCC also has a number of other commands and organizations promoting security and peaceful development in the Caribbean-Central America region, such as the Joint Interagency Task Force-South (counterdrug).

UNITED STATES CENTRAL COMMAND

2-36. The United States Central Command (USCENTCOM) GCC's lead subcombatant command for SOF missions is the Special Operations Component, United States Central Command (SOCCENT). When deployed forward, SOCCENT is referred to as the Combined Forces Special Operations Component Command. The Combined Forces Special Operations Component Command commander has a standing JSOTF operating in Iraq and Afghanistan. USCENTCOM conducts numerous planning conferences with its components. These conferences are promulgated by various Chairman of the Joint Chiefs of Staff and USCENTCOM planning orders. The ASCC for the USCENTCOM GCC is the United States Army, Central Command (USARCENT), with Title 10 responsibility as the lead Service providing common user/items support to Army forces deployed in the USCENTCOM AO. USARCENT's senior Army logistics command is the First Theater Sustainment Command (1st TSC). The 1st TSC and its subordinate elements are responsible for executing theater of operations opening, distribution, and sustainment, to include sustainment of deployed SOF.

UNITED STATES EUROPEAN COMMAND

2-37. The United States European Command (USEUCOM) GCC is dual-hatted as he is also the Supreme Allied Commander, Europe (SACEUR), for the North Atlantic Treaty Organization (NATO). It is the only forward-deployed geographic combatant command HQ. The ASCC for the USEUCOM GCC is the U.S. Army European Command. USEUCOM has an extensive AOR with 92 separate countries. The TSOC is the Special Operations Component, United States European Command (SOCEUR). SOCEUR is responsible to the USEUCOM/SACEUR for SOF readiness, targeting, exercises, plans, joint training, NATO/partnership activities, and execution of counterterrorism, peacetime, and contingency operations. The 21st TSC supports USEUCOM and is responsible for executing theater of operations opening, distribution, and sustainment, to include sustainment of deployed SOF.

UNITED STATES AFRICA COMMAND

2-38. The United States Africa Command (USAFRICOM) AOR covers more than 11.7 million square miles, accounting for 20 percent of the earth's land mass. Africa's nations include approximately 900 million people, constituting 14.2 percent of the world's population. More than 400 ethnic groups live in Africa, speaking more than 2,000 languages and dialects and practicing a wide variety of religious traditions. The issues currently impacting USAFRICOM's AOR include terrorism, enduring conflicts, drug trafficking, territorial disputes, illegal immigration, and natural disasters. Although the nation of Egypt is located on the African continent, it maintains its traditional relationship with USCENTCOM. However, USAFRICOM still coordinates with the Egyptian government on security issues relating to the continent's security. The USAFRICOM TSOC, SOCAFRICA, responsibilities mirror those of other GCC TSOCs. These responsibilities include advising the GCC on the capabilities of SOF and coordination, planning, and control of all SOF on the African continent.

LOGISTICS INFORMATION SYSTEMS SUPPORT

2-39. The USASOC DCS G-4, Systems Management Division, Sustainment Automation Support Management Office (SASMO) provides automation and information system support for all ARSOF. No other sustainment automation support capabilities reside within the ARSOF units, to include the 528th SB(SO)(A). The SASMO provides support in sustaining and operating sustainment-focused automation and information systems, while units are conducting operations from CONUS installation sites in preparation for deployments and during reset operations. This support includes the installation, testing, loading, and troubleshooting of sustainment software; operator and direct support maintenance on systems hardware, connectivity, and associated components; and systems management.

2-40. ARSOF units deploy with information systems capable of connecting to the theater of operations logistics infrastructure and receive SASMO support on an area basis by the supporting TSC logistics units. Information on the different information systems can be found in ADRP 4-0.

Chapter 3

Sustainment Brigade

The 528th SB(SO)(A) is unique when compared to other Army sustainment brigades in that it maintains global situational awareness of sustainment support structures. The 528th SB(SO)(A) is focused at the operational level for sustainment planning and synchronization, and is designed to deploy small, modular teams. The 528th SB(SO)(A) can also serve as the senior logistics unit for an ARSOF-led JSOTF. With the right augmentation and growth, it can establish theater of operations opening and ISB operations with tailored multifunctional Army sustainment enablers.

STAFF STRUCTURE

3-1. The brigade staff is organized to plan and synchronize Army-common logistics and sustainment requirements and support for deployed ARSOF through coordination with TSOCs and theater of operations ASCCs. To this end, the 528th SB(SO)(A) staff structure includes regionally focused, forward-stationed ALEs, which reside with the ASCC and the TSOCs. ALEs are small teams of multifunctional logisticians that serve in direct support of TSOC planning efforts, exercises, and operations to ensure that ARSOF employed in the TSOC AOR are properly sustained.

3-2. In addition to the theater of operations-specific ALEs, the brigade HQ at Fort Bragg, North Carolina, includes an Operations Division comprised of a plans section and a SPO section. The SPO section is organized for mission command of the brigade's home station operations center (HSOC) and to task-organize, train, and deploy ASPO cells to synchronize ASCC-provided logistics support to ARSOF units. The plans section reinforces ALE planning efforts by coordinating SOF-peculiar, strategic, and Title 10 support through HQ, USASOC. Figure 3-1, page 3-2, shows the ARSOF tactical-to-strategic planning model.

MISSION

3-3. The mission of the 528th SB(SO)(A) sets the operational-level logistics conditions in order to enable ARSOF missions. The mission-essential tasks include the following:

- Coordinate ARSOF logistics requirements, plans, Army-common logistics and sustainment in six geographic combatant command AORs to support deployed ARSOF and joint SOF where the Army is the executive agent. ALEs accomplish this by working with both the TSOC and ASCC to ensure ARSOF logistics requirements generated by TSOC plans, exercises, and operations are integrated into the ASCC concept of support for the theater of operations.
- Deploy operational-level logistics synchronization capabilities in support of ARSOF-led JSOTFs. The 528th SB(SO)(A) deploys ASPO cells to collocate with TSC/ESCs to synchronize ASCC-provided logistics support and SOF-peculiar logistics support to ARSOF units.
- Provide expeditionary initial resuscitation and stabilization, limited holding, critical care patient staging, and en route critical care to deployed ARSOF. The special operations resuscitative team can provide reinforcing support and integrate with forward surgical teams.
- Train, resource, and equip the 112th Signal Battalion (Special Operations) (Airborne).
- Deploy a tailored brigade HQ for mission command of operational-level logistics in support of ARSOF missions, until relieved by ASCC logistics mission command capabilities. The 528th SB(SO)(A) is capable of providing mission command of Army CSSBs operating in support of ARSOF for up to 6 months.

Chapter 3

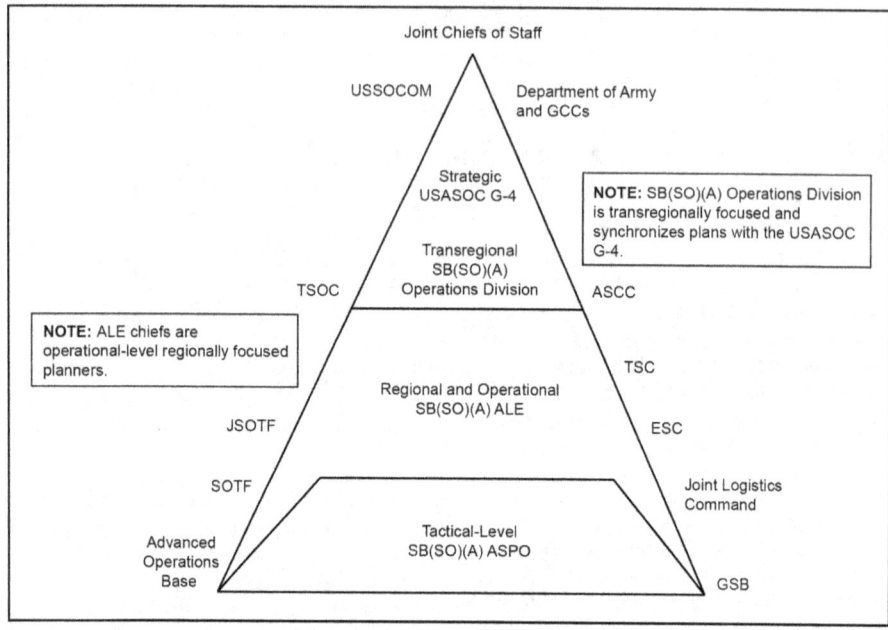

Figure 3-1. ARSOF tactical-to-strategic planning model

CONCEPT OF EMPLOYMENT

3-4. The 528th SB(SO)(A) ALEs are permanently employed in their specific region by being stationed with or in close proximity to each TSOC and theater of operations ASCC where they coordinate ARSOF logistics requirements, plans, and Army-common support in the GCC's AOR to support deployed Army and joint SOF when the ASCC provides support to other nations and services.

3-5. During initial-entry operations, the 528th SB(SO)(A) Operations Division reinforces ALE planning efforts from its Fort Bragg-based HSOC. The Operations Division may also reinforce ALE efforts in the region by locating Operations Division personnel forward with the ALE, when required.

3-6. During initial-entry operations, the 528th SB(SO)(A) may be directed to deploy a tailored brigade HQ for mission command of operational-level logistics in support of ARSOF missions The brigade is not resourced to operate as a stand-alone HQ on account of the lack of base-operations enablers. The brigade requires conventional force augmentation or activation of its USAR companies to perform this mission. During initial-entry operations, the brigade—in addition to performing specific reception, staging, onward movement, and integration (RSOI) functions—may coordinate for a theater of operations-level stockage base and direct logistics support to units deployed forward into their AOs. As the theater of operations grows and matures, this sustainment function will transition on order to an Army sustainment brigade tasked to provide theater of operations distribution and/or to an operational-level sustainment brigade in the theater of operations. This tailored logistical mission command node is for a limited duration until Army sustainment brigades can deploy and assume control of logistics under the ESC.

3-7. If Army conventional force TSC/ESC units are currently in the theater of operations, the 528th SB(SO)(A) may deploy an ASPO cell to collocate with the TSC/ESC, JSOTF HQ, GSB, or RSOD to synchronize ASCC-provided logistics support to ARSOF units.

Sustainment Brigade

3-8. The 528th SB(SO)(A) Operations Division maintains an HSOC in CONUS manned by the Operations Division's SPO section as a reachback capability that provides coordination with the Operations Division's Plans Section and USASOC staff for further analysis in support of the ASPO and ALE missions. The Operations Division is transregionally focused and collaborates daily with the regionally focused forward-stationed ALEs and forward-deployed ASPO cells in order to maintain global visibility of logistics networks. The Operations Division provides logistics planning and analysis and provides the ALEs and ASPO cells a link to the CONUS Army sustainment base.

3-9. The aligned ARNG Soldiers and equipment from the Special Troops Company and forward support company can be mobilized or tasked (through annual training and so on) to support the 528th SB(SO)(A). When employed in support of the brigade, the ARNG Soldiers provide the base operating support capabilities, such as engineering, base operations, food service and field feeding, communications, maintenance, unit ministry team (UMT), staff augmentation for personnel, logistics automation management office, aerial delivery capability, and Role 2 HSS. The ARNG forward support company is designed to execute tactical-level logistics operations as directed by the 528th SB(SO)(A).

ORGANIZATION

3-10. The 528th SB(SO)(A) is made up of Active Army and ARNG units. Figure 3-2 shows the brigade organization.

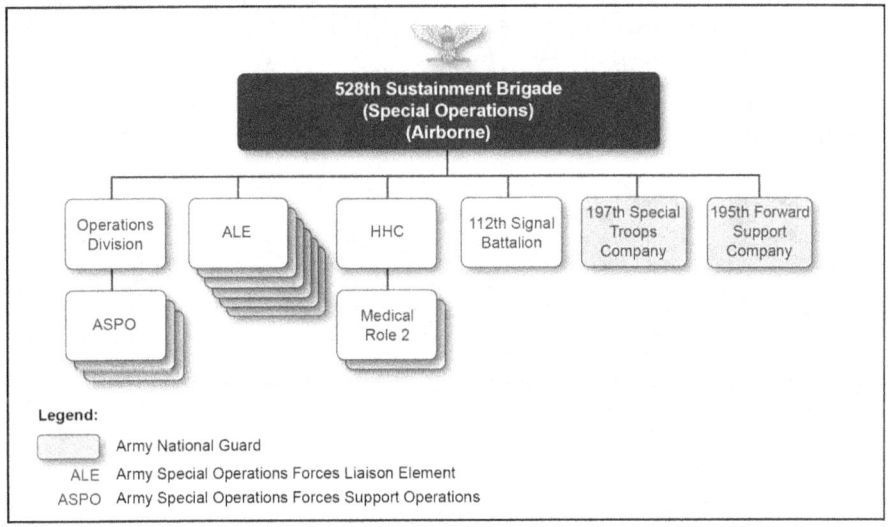

Figure 3-2. 528th Sustainment Brigade (Special Operations) (Airborne)

ARMY SPECIAL OPERATIONS FORCES LIAISON ELEMENTS

3-11. ALEs are the logistics service and support planning and coordination link between the GCC, TSOC, 528th SB(SO)(A), and the ARSOF command structure (SFODA to JSOTF level). The 528th SB(SO)(A) ALEs are not a source of supplies, funds, or augmentation personnel for logistics support. On occasions when the ASCC is not involved in the operation or deployment of ARSOF, ALEs may be tasked to coordinate directly with foreign vendors, U.S. Embassies, and allied forces.

3-12. The ALE consists of one lieutenant colonel (MOS [military occupational specialty] 90A), one major (MOS 90A), one master sergeant (MOS 92A), and three sergeants first class (MOS 88N, 91X, and 89B), who assist in the integration of ARSOF support units with theater of operations assets. It also—

- Serves as the primary interface between the ASCC, TSOC, and USASOC staffs on logistics and health support issues.
- Maintains situational awareness on ARSOF units forecasted and apportioned to or currently operating in the theater of operations.
- Works with the TSOC J-4s to establish and maintain a theater of operations logistics estimate to assist ARSOF units with operational-level planning for missions in the theater of operations.
- Plans, coordinates, and ensures theater of operations Army logistics and theater of operations-opening units facilitate ARSOF RSOI requirements.
- Attends planning conferences at the TSOC, ASCC, USASOC, and 528th SB(SO)(A).
- Given specific TSOC missions, develops an ARSOF concept of support with recommended specified tasks to ARSOF units.
- Provides ARSOF with information concerning Army support in theater of operations-named operations, exercises, and training events.
- Provides TSOCs and ASCCs information concerning ARSOF logistics requirements to support theater of operations-named operations, exercises, and training events.
- In conjunction with the 112th Signal Battalion (Special Operations) (Airborne), coordinates communications support to ARSOF.

3-13. The 528th SB(SO)(A) employs ALEs to support the full spectrum of ARSOF missions. To realize its full capabilities, the ALE requires advanced planning and knowledge of the supported ARSOF unit before deployment. ALEs allow the unit commander and unit-level logisticians to focus their attention on internal and tactical support matters, while the ALE coordinates assets external to the unit. Direct coordination with this element is encouraged.

OPERATIONS DIVISION

3-14. The Operations Division controls the plans and SPO sections. The Operations Division exercises directive authority over subordinate sustainment brigade units during the performance of current sustainment and support operations. The Operations Division, through the use of interoperable automation and communications, performs the daily management functions associated with tasking control for external support operations. It becomes a "fusion information center" to collect, analyze, and anticipate logistic requirements in order to direct actions to better support ARSOF. The SPO section and assigned ASPO cells, in coordination with the ALEs and USASOC staff, provide the global logistical common operating picture (LCOP) of ARSOF logistic and sustainment operations that enable synchronization of logistic, signal, health service, and personnel services support.

Plans Section

3-15. The plans section of the Operations Division is transregionally focused at the operational level of logistics and serves as the link between the ALEs, USASOC, and USASOC units in the deployment and sustainment planning process. The plans section assists in the development and publishing of logistics concepts of support for operations and exercises in coordination with the ALE. It coordinates with the ALE and SPO section on a continual basis to ensure that logistics requirements are properly articulated to the USASOC. This section coordinates with the theater of operations planners in the USASOC G-3 and G-4 to recommend task organization of units and logistics structures.

Support Operations Section

3-16. The SPO section synchronizes global logistical support for ARSOF. The SPO section provides supervision for three ASPO cells, transportation section, supply and services section, ammunition section, and the maintenance management section. The SPO section is responsible for manning the HSOC and providing a reachback capability for deployed ARSOF, ALEs, and ASPO cells. This section maintains global visibility on ARSOF logistical support through daily contact with the ALEs and ASPO cells. The SPO section provides logistical analysis and acts as an entry point for CONUS-based support systems.

3-17. The SPO section maintains communication with the U.S. Army SF Command regarding upcoming GSB deployments in order to assist with relief in place, transfer of authority (RIPTOA) and provide continuity on logistics operations within deployed theater of operations. The SPO section is also responsible for answering requests for forces for brigade assets when requested by GSBs, SOTFs, or deploying/deployed assets.

Army Special Operations Forces Support Operations Cells

3-18. The ASPO cells coordinate, monitor, and synchronize logistics support for JSOTF operations, for other ARSOF missions, and for joint SOF where the Army is the executive agent. While ALEs are focused on logistics planning and coordination at the strategic end of the operational level of logistics, ASPO cells are focused at the tactical end of the operational level of logistics. The ASPO cells deploy from the 528th SB(SO)(A) into the JOA (Figure 3-3, page 3-6) in one of the following four employment options:

- To reinforce an SFG staff for operations in logistically immature theater of operations.
- To coordinate and monitor Army-common and SOF-peculiar logistics and Army Health System (AHS) support for ARSOF by collocating with deployed ESCs and TSCs.
- To provide JTF/JSOTFs with logistics planning and coordination capability.
- With augmentation from the 528th SB(SO)(A), to provide early-entry logistics mission command capability in support of an ARSOF-led JTF until a theater of operations logistics infrastructure can be developed.

When not deployed, the ASPO cells man the SPO section in the HSOC and provide reachback for the ALEs.

Transportation Section

3-19. The transportation section monitors the overall total asset visibility/in-transit visibility of all commodities and unit movements when forward-deployed. The transportation section plans for the use of aircraft and surface transportation assets specifically allocated or attached for logistics and distribution missions. It can coordinate the consolidated shipments of materiel, monitor all inbound and outbound clearances, and coordinate for movement control team support, as appropriate. It maintains automated transportation movement control, tracking, and request systems. All other sections will channel information to this section to improve the total distribution "pipeline" visibility and to allow for overall coordination, prioritization, and decisionmaking to be made by the 528th SB(SO)(A).

Maintenance Section

3-20. This section maintains the Army equipment status reporting data and assists with providing integrated, automated maintenance management for armament-combat vehicles, automotive ground-support equipment, and communications-electronics equipment. This section plans and forecasts maintenance and related material requirements based on future operational plans. It supervises the preparation and maintenance of inventory and activity reports, and recommends cross-leveling and evacuation of the maintenance workload when deployed in support of ARSOF. This section maintains the automated, integrated maintenance information system for the 528th SB(SO)(A) and advises the brigade commander on maintenance and readiness issues.

Chapter 3

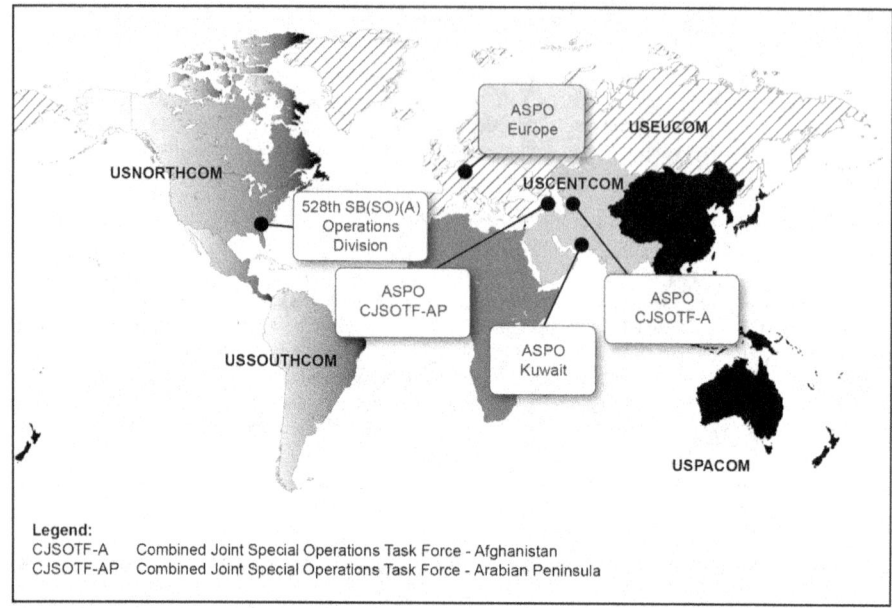

Figure 3-3. Example of ASPO cell deployments

Class V Section

3-21. When deployed, the Class V section facilitates supply management of all ammunition operations within its AOR or JOA. The section maintains asset visibility of all munitions, and assists in required supply rate/controlled supply rate analysis, maintaining the munitions LCOP through total asset visibility/in-transit visibility, managing munitions flow within the AOR/JOA, and providing the conduit to national-level providers.

Supply and Services Section

3-22. The supply and services section assists with forecasting, coordinating, and supervising supply management for water; Classes I through IV, VII, and IX supplies; mortuary affairs; and field services for ARSOF. This section assists in determining the requirements and recommends priorities for the allocation and distribution of supplies. It maintains visibility of on-hand and inbound supply stockages, recommends authorized stockage list (ASL) mobility requirements, and develops basic load recommendations. In addition, this office provides advice on the receipt, storage, and distribution of supplies within its AOR or JOA.

Special Operations Resuscitative Teams

3-23. The medical section provides the only active expeditionary medical capability in USASOC. The primary function of the special operations resuscitative team is the integration of surgical assets (Army forward surgical teams, United States Air Force [USAF] SO surgical teams, or joint forces surgical assets) in austere environments. Each team is capable of providing sick call, advanced trauma management, limited lab, blood storage, X-ray, patient administration, and critical care patient-hold services. Each team can hold up to four critical patients and six noncritical patients and is capable of accepting surgical assets to provide resuscitative surgical intervention in an austere environment or in an environment where general purpose medical assets are unavailable. All 528th SB(SO)(A) AHS assets are modular in nature and can be tailored to meet the specific needs of the supported SOF unit.

112th Signal Battalion (Special Operations) (Airborne)

3-24. The mission of the 112th Signal Battalion (Special Operations) (Airborne) is to provide operational and tactical communications for joint and Army SO commanders in support of contingency and crisis-action operations, and global operations against terrorist networks. The signal battalion can provide signal force packages in support of ARSOF and other organizations, as directed. Additional tasks include providing communications for standing JTFs in USASOC and USSOCOM, as well as being the component subordinate unit to manage all USASOC satellite systems. The signal battalion also provides required coordination between the five theater of operations network operations and security centers, Defense Information Systems Agency, TSOC J-6s, and USSOCOM.

3-25. The 528th SB(SO)(A) is the brigade-level HQ for the 112th Signal Battalion (Special Operations) (Airborne). The signal battalion provides signal support and limited direct service support to ARSOF. The signal battalion also plans, engineers, monitors, sustains, and maintains ARSOF communications elements. FM 3-05.160, *Army Special Operations Forces Communications System*, provides further information on ARSOF communication capabilities.

SPECIAL TROOPS COMPANY

3-26. The Special Troops Company is an airborne ARNG MTOE organization designed to augment and round out the 528th SB(SO)(A) to provide staff and base operating support. It is available for deployment worldwide in support of contingency missions. The Special Troops Company has the capabilities of maintenance, field feeding, medical (special operations resuscitative team), mortuary affairs, laundry and bath, limited base support, and aerial delivery capability. The unit is also designed to augment the 528th SB(SO)(A) staff.

FORWARD SUPPORT COMPANY

3-27. This multifunctional, airborne forward support company provides limited warehousing, Class III, Class V (ammunition transfer point), water, food service, maintenance (including SOF-peculiar equipment), transportation, limited engineer, and limited medical support to deployed ARSOF units. The forward support company is an ARNG company under the command of the brigade troops battalion commander and available for deployment worldwide in support of contingency missions.

This page intentionally left blank.

Chapter 4

Special Forces Group

Organic sustainment capabilities of the SFG(A) are less than comparable to Army units by design, so SF units rely heavily upon regional or geographic combatant command theater of operations support capabilities for support beyond their organic capabilities. The planning and execution of sustainment support for an SF detachment must be nested within the GCC's concept of support and tailored to integrate with the nation and region in which they are operating.

OVERVIEW OF SPECIAL FORCES SUSTAINMENT

4-1. SF detachments may find themselves with emerging requirements that require rapid response and exceed their combat load. The operational effectiveness of deployed forces cannot be achieved without being enabled by Service partners, other agencies, and, in some cases, by the indigenous forces. SF, as well as other SOF components, are the cumulative product of Service-provided personnel and Service-common equipment with specialized training and SO-peculiar modifications. This is true when it comes to the sustainment warfighting function and its major subfunctions—logistics, personnel services, and health services. To be effective, in most theaters of operation, SF elements must understand ARSOF organic sustainment capability as well as other ARSOF enablers, to include Army and joint capabilities. They must understand the basic principles and imperatives of ARSOF logistics and the other subfunctions of sustainment.

4-2. On every deployment, an SF detachment should always ask who is providing their support. For the most part, there is no habitual relationship of theater support units with their SF customers, with an exception being limited organic capability. The Service components do not perceive SOF support as a core Service function, and SF support consequently has a low priority for resources. Early coordination and mutual support between all the players is imperative. Some of the risks and challenges ARSOF have are due to limited logistics and sustainment capability. ARSOF must rely upon the regional or GCC theater infrastructure or the HN for support above the organic level. In some cases (TSCP event or joint combined exchange training [JCET] events), small and remote SF units may require extra sustainment support from conventional forces. Stockpiling supplies and equipment or reaching back to home station may be necessary even though this may not be a generally accepted practice.

4-3. Because of the nature of an SF detachment's missions, sustainment (logistics) planning considerations differ somewhat from what a conventional Army element would consider. SF detachments need to be able to plan for a variety of operations, to include humanitarian, civil, and security assistance programs. Deployed SF are usually in isolated and austere locations where distribution and resupply are key considerations that may require ARSOF-unique capabilities and/or the total dependence on indigenous forces and HNS. Sustainment planning must first consider the existing infrastructure in the JOA and the country or region in which they operate from or within.

GROUP SUPPORT BATTALION

4-4. The GSB is the primary logistics provider in the SFG(A). Its mission is to plan, coordinate, synchronize, and execute logistics operations in support of the SFG(A) and when it is acting as the JSOTF. When ASCC logistics support is unavailable or not established in the JOA, the GSB will be the primary CUL provider. The GSB commander is the senior logistician in the SFG(A) and advises the group commander in logistics management and execution. Much like the SF warrior, the GSB logistician is a dedicated professional whose primary focus is "sustaining the SOF warrior." The GSB is a joint and

Chapter 4

multinational-capable organization in that it can accept, integrate, and employ augmentation of assets from other Services and nations. In a JSOTF model, the GSB commander may assume the role of deputy commander for support in addition to commanding the battalion. As the deputy commander for support, the GSB commander develops the concept of support for the JSOTF, including elements from the Marine Corps Special Operations Command, Naval Special Warfare Command, and Air Force Special Operations Command.

CAPABILITIES

4-5. The GSB provides rapidly deployable multifunctional logistics and HSS/FHP. In developed theaters of operations, the GSB synchronizes its support with the conventional forces. A JSOTF may operate across entire countries or in multiple countries based on the operational construct. No single support battalion can cover this geographical dispersion. The GSB coordinates with conventional force sustainment brigades, the ESC, and the TSC to enable an "area support" concept in support of each SOTF, advanced operations base, and SFODA. Area support enables SOF elements in the vicinity of conventional force bases to receive general sustainment support. The 528th SB(SO)(A) provides a key link to the conventional forces via embedded ALEs and ASPO teams in tactical and strategic conventional force sustainment structure.

4-6. The GSB and forward support companies may require Army logistics augmentation to provide logistics support during sustained operations, or for a capability not organic to the SFG(A). This augmentation may be necessary when the SOTFs are set up in undeveloped theaters without established Army theater opening, theater distribution, or area support; when SOTF bases are not established at fixed facilities; or when a high percentage of SF operational detachments are committed simultaneously.

4-7. The GSB is often required to conduct split-based operations outside the continental United States (OCONUS) while it simultaneously manages and executes critical logistics functions in the garrison environment, such as the supply support activity (SSA), weapons and electronic maintenance, operations detachment, strategic mobility, and consolidated aerial delivery and rigging operations. These garrison functions must continuously operate in order to support elements of the SFG(A) that are performing training and exercises (such as premission training) and TSCP events (such as JCET events). Thus, when deployed, the GSB is a geographically dispersed split-based operation with elements supporting garrison training, exercises and sustainment, and elements forward-deployed in support of the SFG(A) and JSOTF operation.

ORGANIZATION

4-8. The GSB controls consolidated logistical facilities at the JSOTF and projects sustainment operations by ground and air assets. The headquarters and headquarters company of the GSB provides organic battalion-level administrative and supply support for all assigned and attached personnel and coordinates external support for the SF battalions through the SPO section. Figure 4-1, page 4-3, shows the GSB organization.

Support Operations Section

4-9. The SPO section within the GSB is the hub of multifunctional logistic operations in support of the SFG(A). It conducts continuous logistics preparation of the battlefield; develops and synchronizes the overall concept of support; plans and coordinates for ground/air resupply; plans for landing zones in the vicinity of the SOTF; and develops logistics synchronization matrices. The SPO section plans and recommends the allocation of resources in conjunction with the supported chain of command.

4-10. The SPO staff is made up of operations, intelligence, movement, electronic warfare, procurement, petroleum, ammunition, supply, maintenance, food service, and mobility (to include strategic air and ground) subject-matter experts. The GSB SPO section's number one priority is to provide support to each SF line battalion and attached elements. Each battalion S-4 and service detachment commander will interact with the SPO section whether in CONUS or deployed. This SPO section, under the direction of the SPO officer, provides centralized, integrated, and automated command, control, and planning for logistical management operations within the SFG(A). It coordinates with logistics operators and medical personnel in the fields of supply, maintenance, FHP, ammunition, aerial resupply, and movement management for the support of units assigned or attached in the SFG(A)/JSOTF area. The SPO officer and his sections coordinate directly with the supported unit's support center (SPTCEN) and S-4 to synchronize logistics support. The GSB SPO section directly affects SFG(A)-level logistics through management of the group maintenance, mobility, SSA, and rigger warrant officers aligned within the section. Open communication between the GSB SPO section, SFG(A) executive officer, S-3/J-3, and S-4/J-4 is imperative for seamless support execution.

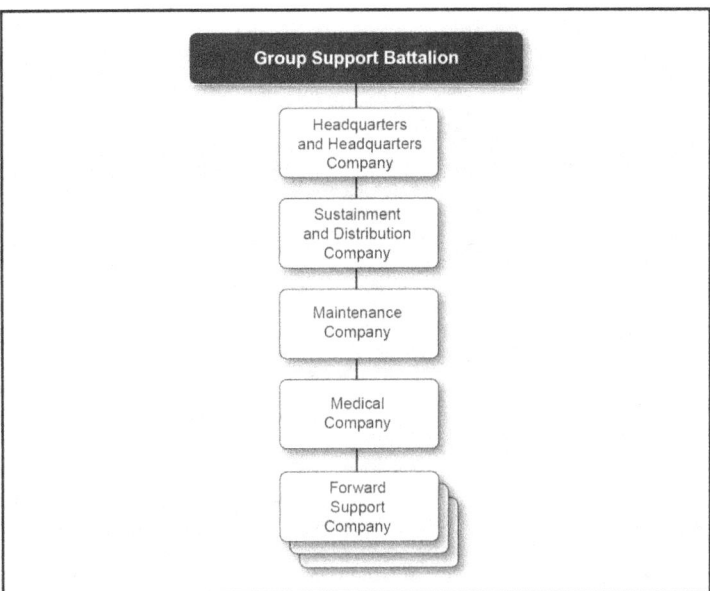

Figure 4-1. Group support battalion

4-11. During early-entry operations, the GSB SPO section may be augmented with the 528th SB(SO)(A) ASPO liaisons to enhance the SPO section's capabilities, enabling it to perform limited theater-opening and theater-distribution functions. TSC elements may also augment the GSB with conventional force logistical assets required to conduct theater-opening and theater-distribution area support functions within a SOF theater.

Support Operations Transportation Section

4-12. The SPO transportation section coordinates and monitors transportation movements (air and ground) of personnel and replenishment stocks. The transportation section has two traffic management coordinators—a mobility warrant officer (MOS 882A) and senior movement noncommissioned officer (NCO) (MOS 88N) assigned to control and synchronize the movement operations for the SFG(A). The traffic management coordinators monitor, control, and supervise the movement of personnel, equipment, and cargo by air, rail, highway, and water. They determine the most efficient mode of transport that accomplishes mission requirements. Specific functions of the traffic management coordinators within the GSB are to supervise cargo documentation and movement control for all transportation modes. They ensure allocation of transport capability is appropriate to accomplish each mission in a cost-effective manner. The GSB SPO transportation section is the hub of strategic movement information before, during, and after deployment operations and readily provides real-time updates to the supported units.

Chapter 4

4-13. The traffic management coordinators provide and use in-transit visibility systems in the ARSOF JOA. The GSB movement control NCO will coordinate the logistics information systems to develop inbound and outbound requirements. The duties of the SPO movement team include the following:
- Validate battalion movement plans and annexes in support of logistics or concept plans.
- Resolve movement priority conflicts in coordination with the SPO officer and SFG(A) S-3.
- Coordinate subordinate unit movement requirements with echelons above brigade.
- Regulate main supply routes used for unit moves.
- Operate the Movement Tracking System.
- Coordinate intratheater movement of cargo, personnel, and aerial resupply.

S-4 Section

4-14. The SFG(A)'s S-4 section has the critical role of identifying all logistics support requirements necessary to achieve mission success. Upon receipt of mission orders, the group S-4, along with all key group and battalion logistics staff, will assist the commander in acquiring the required support and services from internal assets (SPTCEN, GSB SPO, and service detachment). During garrison operations, the group S-4 operates much like a typical brigade S-4 by providing the commander with logistics and supply policy, property acquisition, and accountability, as well as oversight of subordinate battalions' Command Supply Discipline Program compliance. While deployed to a named operation, the group S-4 may assume the role of JSOTF J-4. As such, his duties include, but are not limited to, the following:
- Establishing supply policy.
- Preparing operational needs statements.
- Requesting HNS and SORs.
- Supervising the group issue point, the force modernization/protection office, the contracting office, and the property book officer (PBO).
- Validating all group logistics requirements.

4-15. The SFG(A) S-4 and the GSB SPO must collaborate early on to avoid conflict and confusion among those they support. The S-4 and GSB SPO section must be fully integrated in order to support the complex mission set. The S-4 should be focused on the internal support requirements and readiness of the SFG(A). The GSB SPO is the single logistics integrator between the GSB, SF battalions, and the theater sustainment providers. The GSB commander, SPO section, and S-4 are the key logistics sustainment leaders in the SFG(A).

Sustainment and Distribution Company

4-16. The sustainment and distribution company is a multifunctional logistics company consisting of the supply, service, distribution, and airdrop support platoons. The company provides maintenance, Classes I through IX supplies, water production, bare-base support, aerial delivery, ammunition holding, HSS/FHP support, and transportation. The sustainment and distribution company is independently deployable and capable of providing CUL support to a force package of approximately 2,200 personnel when combined with the logistics support capabilities resident within the SF battalions. For support to progressively larger-sized SO force packages, in multiple locations, the battalion will depend upon augmentation from the theater Army's TSC.

Supply Platoon

4-17. The supply platoon provides essential sustainment to the SFG(A) units. It provides Classes I, II, III, IV, V, VI, VII, and IX support to the SFG(A). The supply platoon receives, stores, and issues Classes II through VII and IX supplies. It receives and distributes, in conjunction with the transportation platoon, Classes I and VI supplies from the field-ration issue point. The supply platoon also receives and issues Class VII supplies, as required. It maintains limited Class II, III, IV and IX ASLs for the SFG(A). The ammunition transfer and holding point section supports the SFG(A) with Class V supplies and operates the SFG(A)'s holding point. Appendix C outlines the classes and subclasses of supply.

4-18. The SSA is located in the GSB of each SFG(A). The SSA is primarily the central receiving point for all classes of supplies and contracted items, and fielding of SOF-unique and Army-common equipment for each group. The SSA receives, stores, and issues Class I, II, IV, VII, VIII, and IX supplies. It operates the standard Army retail supply support (SARSS-1) and provides connectivity distribution hubs that form a sustainment network. These connections are made up of commercial very small aperture terminals for long-haul communications, coupled with Combat-Service-Support Automated Information Systems Interface (CAISI) wireless equipment to provide local area network connectivity. The SSA also provides necessary repair parts and resupply repairable exchanges. This element uses SARSS-1 and related automated systems to provide ASL stock control, receipt, storage, and issue management, and is capable of receiving but not transporting containerized configured loads.

Service Platoon

4-19. The service platoon provides fresh water and bulk fuel services for the SFG(A). It consists of the water section—which provides water purification, bulk water storage, and distribution—and a fuel section, which receives, stores, and distributes class III(B) fuel. Fuel quality-control personnel reside within the platoon HQ. The water section operates one water purification point with two organic reverse osmosis water purification units (ROWPUs). The two ROWPUs are programmed to be replaced by one 1,500-gallons-per-hour tactical water purification system and lightweight water purifier systems. The fuel section consists of three 2,500-gallon Heavy Expanded Mobility Tactical Truck (HEMTT) fuel tankers, and one 60,000-gallon fuel system supply point. The fuel section also has the capability of establishing a forward arming and refueling point (FARP). The HEMTT fuel tankers are the primary retail vehicles supporting the SFG(A). Generally, the fuel supply point remains in the JSOTF area, while the HEMTT fuel tankers conduct retail fuel support to the SFG(A) in the JOA. The fuel supply point receives fuel from military or commercial vehicles, and issues bulk fuel to the HEMTT fuel tankers.

Distribution Platoon

4-20. The distribution platoon consists of the transportation squad and the movement control team. This multifunctional platoon provides limited transportation supporting the JSOTF distribution missions and arrival/departure airfield control group mission support (capable of 125 short tons line haul or 90 short tons local, with lift for 120 passengers). The movement control team provides coordinating instructions between the USAF tanker/airlift control center and the deploying or redeploying units. The movement control team—

- Validates ARSOF aircraft load plans.
- Inspects and certifies hazardous materials for air movement.
- Requests aircraft and plans aircraft configurations.
- Prepares the movement timetable for passengers and equipment in coordination with battalion-level movement planners.
- Acts as military custom inspectors during arrival/departure airfield control group operations in support of SOTF operations.

Airborne Support Platoon

4-21. The airborne support platoon consists of three sections: aerial delivery operations, personnel pack, and maintenance. This platoon operates in a consolidated facility supporting the entire SFG(A). The platoon is capable of supervising unit preparation of up to ten tons of general supplies and equipment per day for aerial resupply loads up to 2,000 pounds. It provides personnel parachute packing to ARSOF and joint elements and provides all levels of maintenance of aerial delivery items, to include ancillary oxygen equipment used for military free-fall operations.

Maintenance Company

4-22. The maintenance company consists of the following sections: HQ, maintenance control, ground maintenance, electronic maintenance, armament maintenance, and base support. It provides base operations support and field-level maintenance for Army-common and select SOF-unique automotive, power

Chapter 4

generation, armament, construction, quartermaster, communication, electronic, and ground support equipment. The diverse SO-peculiar equipment density requires on-site contractor maintenance coordinated through the battalion/group S-4 section. The maintenance control section provides maintenance information management to the SPO section by transmission of data and wireless transmission by CAISI, very small aperture terminals, and other communication systems, from the maintenance control section's Standard Army Maintenance System-Enhanced (SAMS-E) 1 box to the GSB SPO section's SAMS-E2 box. When the SAMS-E is not available, the maintenance control section will use other methods to transfer data.

Medical Company

4-23. The medical company consists of the HQ, health service skills, ancillary service skills, and HSS sections. The company provides Role 1 care to assigned and supported personnel. Its HSS/FHP capabilities can be increased with augmentation from conventional AHS organizations. The medical platoon and battalion medical section has an assigned medical officer, dental officer, veterinary officer, physical therapist, physician assistant, medical logistics officer, medical operations officer, and environmental science officer. In addition, it contains noncommissioned officers and enlisted personnel trained in preventive medicine, medical logistics, dental care, health care, and biomedical equipment repair. This provides the medical platoon with greater medical capabilities than those assigned to conventional Role 1 AHS organizations.

Forward Support Company

4-24. There are three organic forward support companies within the GSB; they are designed to provide support to the SF battalions. The forward support company consists of the sustainment, distribution, and maintenance platoons. The forward support company is a multifunctional logistics company providing maintenance, limited Classes I through IX supplies, fuel and water production, ammunition holding, and transportation. The forward support company is independently deployable and capable of providing for the entire SF battalion and its attached elements. When the SF battalion establishes a SOTF, the forward support company commander may coordinate and supervise the SPTCEN logistics activities. The forward support company—

- Provides continuous battle tracking.
- Assists with developing the concept of support for the battalion OPORD.
- Conducts tactical and logistical coordination with higher, adjacent, and supported units, as appropriate.
- Oversees the development of the daily logistics packages by the service detachment supply section and the company supply sergeants.

Sustainment Platoon

4-25. The sustainment platoon provides essential sustainment to the SF battalion. It provides Classes I through IX supplies (very limited II, IV, and VII), and this capability is nested in the GSB. It receives and distributes, in conjunction with the distribution platoon, Classes I and VI supplies from the field-ration issue point. The ammunition transfer and holding point section supports the SF company with Class V supplies and operates the battalion's holding point.

Distribution Platoon

4-26. The distribution platoon is made up of the transportation squads and movement control teams. This multifunctional platoon provides limited transportation supporting the SOTF distribution missions and the arrival/departure airfield control group mission. The movement control team provides coordinating instructions between the USAF tanker/airlift control center and the deploying or redeploying units. The movement control team—

- Validates ARSOF aircraft load plans.
- Inspects and certifies hazardous materials for air movement.

- Requests aircraft and plans aircraft configurations.
- Prepares the movement timetable for passengers and equipment in coordination with battalion-level movement planners.
- Acts as military custom inspectors during arrival/departure airfield control group operations in support of SOTF operations.

Maintenance Platoon

4-27. The maintenance platoon provides very limited base operations support and field-level maintenance for Army-common and select SOF-unique automotive, power generation, armament, construction, communication, electronic, and ground support equipment. The diverse SO-peculiar equipment density requires on-site contractor maintenance coordinated through the battalion/group S-4 section. The maintenance control section provides maintenance information management to the SPO section by transmission of data and wireless transmission by CAISI, very small aperture terminals, and other communication systems, from the maintenance control section's SAMS-E1 box to the GSB SPO section's SAMS-E2 box. When the SAMS-E is not available, the maintenance control section will use other methods to transfer data.

LIMITATIONS OF THE GROUP SUPPORT BATTALION

4-28. The GSB is not designed to provide all of the logistics functions. Factors and limitations to be considered are as follows:

- Urban areas, dense jungles and forests, steep and rugged terrain, and large water obstacles limit movement.
- The GSB has no organic mortuary affairs capability for collection, processing, and evacuation without augmentation.
- Laundry and bath is not organic to the SFG(A); support is provided by the 528th SB(SO)(A) or the TSC.
- Limited financial management.
- Limited capability to reconfigure load. Ammunition from echelons above brigade must be in strategic or operational configured loads.
- No firefighting capability.
- Explosive ordnance disposal is not organic to the SFG(A) and requires augmentation from the ASCC.
- Limited human resources other than its own unit S-1 human resources operations, it relies on the ASCC to provide additional critical wartime personnel support.
- Legal support is limited to the assigned SFG(A); augmentation to support all Judge Advocate General functions is required.
- No organic band support.
- No optical fabrication and blood product management support.
- No organic air medical evacuation (MEDEVAC) support.
- No organic sustainment automation support capability.

SPECIAL FORCES BATTALION

4-29. Sustainment support of the SFODAs begins with the SF battalion. The SF battalion commander must know the capabilities and limitations of his organic sustainment operations. He must ensure that his concept of support is synchronized and integrated with his higher HQ and the theater's sustainment support elements. It all begins with the forward support company commander who must be proficient in the tactical employment and sustainment of his company, as well as the SFODAs.

Chapter 4

BATTALION STAFF LOGISTICS RESPONSIBILITIES

4-30. The SPTCEN anticipates, requests, coordinates, and integrates logistics for the tactical mission. The S-4 section leverages GSB capabilities by providing the SPO section in-depth analysis of the tactical plan and the sustainment requirements inherent to that plan. The S-4 section processes requests for replenishing basic loads of all unit elements and monitors the request of Class I, II, III, IV, V, and VII items. The section monitors the equipment and supply status of all battalion elements, prepares supply forecasts for the unit, and submits these to the SPO section. The S-4 will transmit the logistics situation report electronically to the SFG(A) S-4 (or JSOTF J-4) and GSB SPO section, enabling quick identification of problems and proper allocation of resources. The Force XXI Battle Command Brigade and below (FBCB2) also provides map graphics that portray unit locations, grid coordinates, and terrain features so the SPO section can track maintenance on the battlefield.

PREDEPLOYMENT OPERATIONS

4-31. Synchronized mission analysis is conducted by the GSB SPO section, supported units, 528th SB(SO)(A), TSOC, and identified conventional force support elements. Planning and coordination is also executed with Embassy teams, standing JTFs, and joint agencies; for example, the U.S. Agency for International Development, the joint interagency task force, and the DOS. Standing logistics estimates maintained by each TSOC provide the foundation for logistical mission analysis for a specific AO. Additional augmentation requirements are determined throughout the mission analysis process based on the mission requirement and the GSB and service detachment's capability. There are many sustainment-centered actions that must take place during predeployment; for example, support for premission training, filling critical property shortages, and a 100-percent property inventory and hand-receipt split for deploying equipment. The service detachment, battalion S-4, and PBO should be consulted for specific guidance.

DEPLOYMENT OPERATIONS

4-32. Support for TSCP events—as well as unconventional warfare (UW), foreign internal defense (FID), counterintelligence, and security force assistance—require integration with the unit comptroller and 905th Contingency Contracting Battalion specialists. In addition to the GSB and conventional force sustainment structure and capability, special funding authority must be understood to operate on the current and future battlefield. Programs such as Confidential Military Funding, Commanders' Emergency Response Program (CERP), and the National Defense Authorization Act (NDAA) (Sections 1206 and 1208) must be understood by both the SF community and the logistician as they affect the way the United States sustains its forces as well as its partners.

PERSONNEL SERVICES

4-33. Personnel services for a deployment are broken into four phases: predeployment requirements, deployment personnel actions, redeployment, and postdeployment. Predeployment requirements vary based on the deployment notification timeline. At a minimum, a Soldier's DD Form 93 (Record of Emergency Data) and SGLV 8286 (Servicemembers' Group Life Insurance Election and Certificate) are reviewed and updated. Also, deployment timelines must be provided to the SF unit's S-1 section for accurate accountability. Additional requirements are provided by the SF unit's S-1 section based on the varying degree of each deployment.

4-34. Deployed personnel actions are processed through the SF unit's S-1 section unless otherwise stated. It is mandatory that Soldier accountability be enforced. Once redeployment preparations have started, Soldiers need to keep in mind time-sensitive actions, such as special pays and entitlements, as well as awards or evaluations that may need to be completed.

4-35. Once a Soldier has redeployed, he must complete all reintegration personnel requirements within the SF unit S-1 section's established timelines. Specific actions include reviewing the DD Form 93 and SGLV 8286, personnel tempo, dwell time updates, and personnel record updates. These areas are emphasized since these actions are the foundation for personnel and deployment management.

4-36. For additional human resource information and guidance, units/Soldiers can reference FM 1-0 and the Army Personnel Policy Guidance. The deployment cycle checklist can be an additional tool used to capture the required steps needed for an individual to deploy. Specific personnel deployment questions that are not covered in the above references can be addressed by the SF unit's S-1 section or at the local installation Soldier Readiness Center.

HEALTH SERVICE SUPPORT/FORCE HEALTH PROTECTION

4-37. An HSS/FHP planner from the SFG(A) and/or SF battalion medical sections must be involved as early as possible in the planning process. The HSS/FHP planner must produce a straightforward plan without complication to ensure a continuum of care to the full range of SF operational environments. FM 4-02.43, *Force Health Protection Support for Army Special Operations Forces*, includes more information.

4-38. The SF medical planner must plan for all HSS/FHP functional support areas (Table 4-1, pages 4-9 and 4-10) regardless of the mission or the operational environment since SF units and personnel operate over a wide area and in isolated and austere locations with limited AHS support. HSS/FHP planning for SF assets involves numerous considerations that do not apply to conventional forces.

4-39. The FHP plan must be in sufficient detail to designate specific area support responsibilities provided to SF units. A well-developed HSS/FHP plan within the OPLAN/OPORD ensures all medical functional support is available to support SF personnel when required or requested. The HSS/FHP planner must consider all medical functional areas discussed in ATTP 4-02, *Army Health System*, when developing the HSS/FHP plan. Direct support from conventional force medical units may not be available or may be limited during certain operations. Support may also be required for SF units or personnel following the completion of the mission. SF units must coordinate support from theater medical units. To enable this, the SF HSS/FHP planner must develop a comprehensive plan and thoroughly coordinate and update the plan with the supporting medical mission command HQ. In SF operations where medical units are not available in the operational environment, tasking for specific HSS/FHP support requirements and capabilities are critical for adequate HSS/FHP.

Table 4-1. Organic and nonorganic health service support/force health protection

Functional Area	SFODA	SFODB	SF BN	SFG(A)	528th SB(SO)(A)	BCT	MMB	FST	CSH	Theater
Evacuation:										
CASEVAC Platforms (Air/Land/Sea)	X	X	X	X	X	X	X	X	X	X
MEDEVAC (Air)										X
Treatment:										
Sick Call/ Trauma	X	X	X	X	X	X	X	X	X	X
Surgery								X	X	X
Hospitalization									X	X
Patient Holding	Limited	Limited	Limited	Limited	Limited	Limited	Limited	Limited	X	X
Medical Logistics				Limited	Limited	X	X	X	X	X
Preventive Medicine			X	X	Limited	X	X	X	X	X
Veterinary Services					Limited	X				X
Dental Services				Limited	Limited	X	X			X

Table 4-1. Organic and nonorganic health service support/force health protection (continued)

Functional Area	SFODA	SFODB	SF BN	SFG(A)	528th SB(SO)(A)	BCT	MMB	FST	CSH	Theater
Behavioral Health				X					X	X
Laboratory Services					Limited	X	X	Limited	X	X
Medical Information Systems	Limited	Limited	Limited	X	Limited	X	X		X	X

NOTES:
BCT	Brigade Combat Team	FST	Forward Surgical Team
BN	Battalion	MMB	Multifunctional Medical Battalion
CASEVAC	Casualty Evacuation	SFODB	Special Forces Operational Detachment B
CSH	Combat Support Hospital		

4-40. SF elements doctrinally have Role 1 capability, which is the first level of medical care an SF Soldier would receive, and SF elements must depend on area and theater of operations HSS/FHP assets for medical requirements. SF, in coordination with conventional HSS/FHP planners, determine what support is organic and what area and theater of operations assets will be provided by conventional forces. Early coordination and communication is the key to success for HSS/FHP support to SF operations. Figure 4-2 depicts the organic medical personnel structure within an SFG(A). Its capabilities are greatly enhanced beyond those available in conventional Role 1 organizations with its assigned medical, dental, veterinary, physical therapy, environmental science, medical logistics, physician assistant, and medical operations personnel.

SFG(A) Medical Section (HQ)		
Title (MOS/Grade)	Title (MOS/Grade)	Title (MOS/Grade)
Group Surgeon (61N/O-5)	Medical Operations Sergeant (18Z/E-8)	Psychologists (73B/O-4)
Medical Operations Officer (70H/O-3)	Medical Logistics Sergeant (68W/E-6)	
SF Medical Platoon (GSB)		
Physician Assistant (65D/O-3)	Medical Logistics Officer (70K/O-3)	Medical Logistics Specialist (68J/E-6/E-4)
Environmental Science Officer (72D/O-3)	Special Forces Medical Sergeant (18D/E-7/E-6)	Special Operations Combat Medic (68WW1/E-5)
Veterinary Officer (64A/O-3)	Dental Officer (63A/O-3)	Dental Specialist (68E/E-4)
Preventive Medicine Sergeant (68S/E-7)	Physical Therapist (65B/O-3)	
Battalion Medical Section		
Battalion Surgeon (61N/O-4)	Physician Assistant (65D/O-3)	Preventive Medicine Sergeant (68S/E-7)
Special Forces Medical Sergeant (18D/E-7)	Medical Logistic Specialist (68J/E-4)	
SFODB		
	Special Forces Medical Sergeant (18D/E-7)	
SFODA		
	Special Forces Medical Sergeant (18D/E-7/E-6)	

Figure 4-2. Medical personnel structure within the Special Forces group (airborne)

Special Forces Medical Sergeant

4-41. The SF medical sergeant (MOS 18D) forms the backbone of medical care within the SFG(A). There are two SF medical sergeants in each SFODA. SF medical sergeants—

- Provide, train, and advise detachment members, multinational and coalition forces, or indigenous personnel in emergency and routine medical care, emergency dental care, and veterinary care.
- Establish field medical treatment facilities to support detachment operations and prepare the medical portion of area studies, OPLANs, and OPORDs.
- Conduct medical intelligence analysis and prepare health threat and counterthreat briefings and lessons-learned briefings.
- Requisition, assemble, and maintain detachment medical equipment and supplies.
- Supervise routine and emergency medical activities in a field or UW environment.

4-42. USASOC Regulation 350-1, *ARSOF Active Component and Reserve Component Training*, addresses medical sustainment training requirements. Clinical capabilities for the SF medical sergeant are discussed in AR 40-68, *Clinical Quality Management*.

Medical Considerations

4-43. The following characteristics for SF operations must be factored into the plan:

- *Small units and austere FHP capability.* SF unit location (geographical factors, time-distance factors) may require collocation of assets and support on an area basis.
- *Operations in a joint, multinational, and coalition environment.* Operations in these environments require SF medical personnel to have a thorough knowledge of other Service component, multinational, and/or coalition forces' HSS capabilities, limitations, organization, procedures, and national caveats.
- *Remote operating areas and long evacuation routes.* SF elements often operate in areas that impede evacuation by rotary-wing aircraft or where aviation assets are not available. This places a premium on the early application of trauma management and casualty stabilization.
- *Medical evacuation, medical regulating, and casualty tracking.* Medical regulating and casualty tracking requires an understanding of SF missions and the limited availability of replacements. Leaders must account for sensitive equipment and/or documents if the casualty still possesses them when evacuated.

4-44. It is imperative that SF medical sergeants conduct operational medical planning and geographical and environmental medical threat analysis prior to all deployments to determine materiel quantities and specific additional support requirements for that mission. Medics must fully understand the SF principal tasks and the operational, tactical, and geographical constraints associated with those tasks. In planning and coordination for medical support (both internal and external), SF medical personnel must analyze operational mission planning factors against the ten medical functional areas to determine materiel and equipment requirements and develop the mission-specific medical support plan. The ten medical functional areas are medical mission command, medical treatment (organic and area support), MEDEVAC, hospitalization, dental services, preventive medicine services, combat and operational stress control, veterinary services, medical logistics, and medical laboratory services. Once deployed, the SF medical personnel must continually refine the medical support plan by conducting theater medical support analysis and modify the plan as theater medical support assets become available.

4-45. Additional considerations when working with indigenous populations include local medical infrastructure, theater medical rules of engagement (combatant or noncombatant personnel), cultural beliefs, and effects on the population. FM 4-02.43 includes more information on medical planning considerations for SF.

Chapter 4

FINANCIAL MANAGEMENT SUPPORT

4-46. The financial management mission is to ensure that proper financial resources are available to accomplish the mission in accordance with (IAW) commanders' priorities. The primary purpose of financial management is to sustain and support operations until successful mission accomplishment. Financial management is composed of two distinct, but mutually supporting, functions: RM and finance operations. Although independent of one another, these two functions must be integrated into mission planning and execution at every level. Integration facilitates the optimal allocation of financial resources to accomplish the mission. The combined efforts of RM and finance operations work to extend Army forces' operational reach and prolong operational endurance, thereby allowing commanders to accept risk and create opportunities for decisive results.

Resource Management Support

4-47. The RM section is the commander's leading representative responsible for financial management support. RM advises the appropriate allocation and use of scarce resources, to include funding, in the accomplishment of the commander's assigned missions. RM personnel assist commanders by providing a critical capability, which matches legal and appropriate sources of funds with thoroughly vetted and valid requirements. Funding support provides flexibility through nonlethal methods to augment, and in some cases, lead the effort in obtaining the effects the commander is trying to achieve.

4-48. The RM mission is to analyze resource requirements ensuring commanders are aware of existing resource implications in order for them to make resource informed decisions, and then obtain the necessary funding that allows the SF commander to accomplish the overall unit mission. Key RM tasks include the following:

- Providing advice and recommendations to the commander.
- Identifying sources of funds.
- Forecasting, capturing, analyzing, and managing costs.
- Acquiring funds.
- Distributing and controlling funds.
- Tracking costs and obligations.
- Establishing and managing reimbursement processes.
- Establishing and managing the Army Managers' Internal Control Program.

RM personnel also provide a variety of organic support to commanders for overseas contingency operations, Joint Chiefs of Staff exercises, counternarcotics training, and JCET events.

Operational Funds Support

4-49. RM organic support includes use of operational funds (OPFUNDs). USASOC OPFUNDs are governed by USASOC Policy Number 32-09, *Operational Funds (OPFUNDS)*. OPFUNDs can be requested before, during, or after deployments and authorized training events. Units must work with their RM office to request an OPFUND. In general, the commander appoints a pay agent on an additional duty appointment order. This appointment authorizes the pay agent to disburse public currency IAW the special instructions stated in the appointment and the written instructions provided by the financial management commander. The field ordering officer, whom the pay agent supports, receives separate instructions from contracting officials. Field ordering officers and pay agents train and work as a team; the pay agent should participate with their field ordering officer in training and vice versa. The pay agent or field ordering officer may be held personally liable for any payment not IAW the appointment orders or prescribed instructions. The pay agent cannot simultaneously serve as either a certifying officer or field ordering officer. The pay agent uses an official credit or debit card to make payments whenever possible. When it is not possible to use an official credit or debit card to make payments, the pay agent takes the following actions:

- Reviews all Standard Forms 44 (Purchase Order Invoice-Voucher) prepared by the ordering officer.
- Disburses currency for the goods or services as stated on the Standard Form 44, but only after this form has been approved by a field ordering officer.

- Pays for purchases not to exceed established limits. (An agent may not split purchases between two or more vouchers to circumvent the established limit.)
- Clears his account with the disbursing officer that advanced the funds.

Other Funding Support

4-50. Funding support is a complex endeavor and requires RM personnel to leverage multiple appropriations. Some of these appropriations are initially provided for peacetime support, along with appropriations that are newly created by Congress specifically for an operation. Commanders and RM personnel need a thorough understanding of the statutes and regulations that govern the use of appropriated and nonappropriated funding. RM personnel must work closely with the fiscal lawyer to ensure compliance with fiscal requirements established by law. The following discussion highlights the basic appropriations that fund SOF. Multitudes of funding options are available and may include funding sources from other U.S. agencies (for example, intelligence funding, counterdrug funding, and DOS funding). Funding authorities the financial management personnel may leverage before, during, and after contingency operations include the following:

- *Operations and Maintenance, Army (OMA) or Major Force Program (MFP)-2.* SF units receive some direct funding (MFP-2) from Army for some Army-common requirements. SF units use MFP-2 to pay for the day-to-day expenses in garrison, and during exercises, deployments, and military operations. There are threshold dollar limitations for certain types of expenditures, such as purchases of major end items of equipment and construction of permanent facilities. OMA is typically a one-year appropriation and must be obligated in that fiscal year (1 October to 30 September).
- *Operations and Maintenance, Defense (OMD) or MFP-11.* SF units use MFP-11 for training, equipping, and employing SF with SO-peculiar equipment, materials, supplies, and services. MFP-11 is for SOF-unique requirements only. The same dollar limitations apply to MFP-11 as MFP-2 funds.
- *Military Personnel, Army (MPA).* MPA funding is used for pay, allowances, individual clothing, subsistence, interest on deposits, gratuities, and permanent change of station travel (including all expenses for organizational movements) for members of the Regular Army and mobilized Reserve and National Guard Soldiers. MPA funding is generally available for one fiscal year and is centrally managed and funded. Since MPA funding is centrally managed, personnel should plan in advance for the use of MPA funding to ensure receipt in time to satisfy the requirement.
- *Procurement.* While OMA funds day-to-day operations, procurement is typically used for centrally managed items or systems that are considered investment items. These items require the use of procurement funds regardless of cost (or the cost of individual components). Such items can include large pieces of equipment or systems that exceed the expense investment threshold.
- *Research, development, test, and evaluation (RDT&E).* RDT&E funds provide for the development, engineering, design, purchase, fabrication, or modification of end items, weapons, equipment, or materials. This is not an appropriation normally used in the theater by deployed units unless involved in the research, development, acquisition, and testing process. RDT&E funding is available for two years.
- *Military construction.* Military construction provides for the acquisition of land and construction of buildings for which authorizing legislation is required.

4-51. In addition to the funding support provided by the sources listed above, both commander and RM personnel may leverage the additional funding sources listed below (additional information regarding the funding sources listed below is in Chapter 10):

- CERP.
- DOD Rewards Program.
- Emergency and Extraordinary Expense Authority.
- Combatant Commander Initiative Fund.

- Section 1206 Authority.
- Section 1208 Authority.
- Sensitive Mission Fund.
- Memorandum of Agreement.

Finance Operations Support

4-52. Finance operations support is not SF-specific. Finance operations support the sustainment of Army, joint, and multinational operations through the execution of key finance operations tasks. Key finance operations tasks include the following:

- Provide timely commercial vendor and contractual payments, and various pay and disbursing services.
- Oversee and manage the Army's Banking Program.
- Implement financial management policies and guidance prescribed by the Office of the Under Secretary of Defense (Comptroller) and national financial management providers; for example, the U.S. Treasury, Defense Finance and Accounting Service, and Federal Reserve Bank.

LEGAL SUPPORT SERVICES

4-53. Each SFG(A) has one Command Judge Advocate (CJA) position at the grade of O-4, and each SF battalion has a CJA at the grade of O-3. Each SFG(A) also has a paralegal NCO at the grade of E-7, and each SF battalion has a paralegal NCO at the grade of E-6, assigned under the supervision of the CJA. The CJA and paralegal NCOs are assigned as a special staff to each Commander and provide legal advice on the six core legal disciplines listed below (additional information is in Chapter 10):

- Military justice.
- International and operational law.
- Administrative and civil law.
- Contract and fiscal law.
- Claims.
- Legal assistance.

RELIGIOUS SUPPORT

4-54. The SF battalion chaplain has unique capabilities with the freedom of movement and credibility to conduct battlefield circulation from SF companies to SFODA locations to provide religious support. The chaplain does this by nesting his visits with the battalion S-3 and travels to the various units in a systematic way to provide religious support directly to his unit.

4-55. The SF battalion chaplain provides direct religious support to SFODAs, subordinates units, and family members, and may accompany the SFODA on selected missions. He is responsible for providing direct religious support in three ways: nurture the living, care for the wounded, and honor the dead with memorial ceremonies and services. His missions are to advise the commander on all matters of religion, ethics, morals, and morale within the unit.

UNCONVENTIONAL WARFARE SUSTAINMENT

4-56. SF missions require civil support operations be conducted with low visibility and under clandestine conditions. These types of civil support operations are referred to as unconventional logistics. Generally, logisticians providing such support remain cognizant of conventional logistics principles, such as supply chain management, but must adapt existing TTP or develop new TTP to deal with processes such as acquisition, storage, funding, and transportation. Logisticians may use conventional providers within the DOD, but are not restricted to those providers. Demand for such activity can result from missions being conducted across the entire spectrum of military operations.

Special Forces Group

4-57. Unconventional logistics places extraordinary demands on both the human resource system and the Soldiers assigned to conduct these types of civil support operations. Training in areas such as acquisition, storage, commercial and military transportation, and multiple funding streams are often unique and not available within standard Army courses. Unconventional logistics requires innovative and adaptable individuals. These Soldiers must be intensively managed to maintain the needed level of expertise over time while ensuring Soldier promotion and career progression opportunities are maintained despite prolonged service in this area. Experience gained in unconventional logistics, over extended time periods, is invaluable to SF units.

4-58. Sustainment planning for any operation involves the identification of requirements, the provision of organic capabilities to meet those requirements, the subsequent identification of shortfalls, and consideration of leveraged or acquired options to mitigate those shortfalls. This is the conventional logistics estimate model and is applicable for the UW concept of support planning as well. Because of operations security reasons and the complexity of UW mission sets, challenges exist in the unconventional logistics arena that require increased diligence in the planning processes. Some of the key challenges include the following:
- Reduced numbers of sustainers organically assigned to the unconventional logistics sustainment effort because of the time needed to train and grow experienced support personnel.
- The need to conduct the steps in the sustainment lines of effort under covert or clandestine conditions.
- The need for SF elements to perform tasks traditionally handled by supporters.

Note: Training Circular (TC) 18-01, *Special Forces Unconventional Warfare*, Chapter 3, Concept of Employment, contains further information concerning UW sustainment.

PREMISSION CONSIDERATIONS

4-59. The key elements of a good concept of support involve detailed planning with respect to providing goods (supplies and equipment), services, and support (construction and facilities). Conventional operations have established funding streams, supply chains, and players operating support nodes. Conversely, unconventional logistics support often lacks the up-front structure. Once the unconventional logistics requirements have been identified, the unconventional logistics planner must consider how the required goods, services, or support needs will be funded. The unconventional logistics planner will then identify the supply chain or procurement trail and understand that they may be different for each item within the same operation. The identification of players in the operation often includes a METT-TC determination of the point at which a professional sustainer will be involved rather than using an SF element. It could also involve determining what conventional or unconventional systems will be leveraged to provide the required support.

MISSION

4-60. Unconventional logistics must include a plan that adjusts to the branches and sequels of the UW mission. Support requirements sometimes change across the operational phases. The unconventional logistics support planner must anticipate, monitor, and respond to the requirements associated with these changes.

4-61. Unconventional logistics support for UW is different from support to other SF principal tasks. UW missions will require significant quantities of materiel to support resistance forces, specifically guerrillas. The materiel includes lethal and nonlethal aid, some of which may not be organic to the U.S. military supply system. Every effort must be made to maximize the use of indigenous supply sources within the UW operational area. In addition, confiscation, barters or trades, IOUs, donations or levies, and battlefield recovery and purchase is leveraged extensively to minimize demands of external resupply. Understanding the various funding policies that regulate sustainment and partnership functions is required. Reviewing the previously discussed financial management support will provide needed context enabling the execution of various funding programs.

4-62. Establishing an ISB is a critical step in successful UW support. The ISB should have experienced unconventional logistics logisticians, RM personnel, and contracting specialists for rapid acquisition of goods and direct access to nonstandard aviation assets (fixed- and rotary-wing platforms) to enable timely resupply. Supporting UW requires maximum flexibility, financial and contracting agility, and rapid operational and strategic reachback to provide responsive support.

REDEPLOYMENT, REFIT, AND RECOVER

4-63. As with conventional operations, a plan must exist to account for closing current operations, while simultaneously preparing for future missions during redeployment, refit, and recover. Oftentimes, planners do not account for the final step in the life-cycle of sustainment. This step includes the destruction, retrograding, or recollection of issued items IAW regulations, policies, and authorities, and the closing out of contracts. Refitting and recovering equipment and reestablishing stockage levels are paramount to preparing for the next mission. It is important to note that, while the quantity of unconventional logistics sustainers is low, their operating tempo is high, and these sustainers need to reset themselves as well.

4-64. Lessons learned and after action reviews must be captured and shared within the SF community to allow the unconventional logistics planner to leverage the TTP and range of options employed. The use of lessons learned and after action review knowledge sharing must be planned to take place throughout the spectrum of an operation.

TIME

4-65. Time is more of a critical planning consideration for the unconventional logistics sustainment planner than it is for his conventional counterpart. It sometimes takes longer than anticipated for the SF element to receive the required goods and services based on—
- Secure operating conditions.
- Reduced number of sustainers.
- Length of the unconventional logistics mission at large and the time needed to establish and develop relationships.
- The process of obtaining authorities for funding.

4-66. Time is further confounded by the fact that the supply chain often consists of sections that are not necessarily under the control of the unconventional logistics community. Additionally, the goods or services themselves are often low density—such as items to support another nation or SO-peculiar items.

FOREIGN INTERNAL DEFENSE SUSTAINMENT

4-67. When planning sustainment support for a FID mission in a foreign country, the SF sustainment planner must understand the DOD, DOS, and USSOCOM policies and guidance that apply to the specific mission. SF sustainment planners need to base their plans on higher-level guidance, priorities within the GCC/TSOC, and the resources available. In some circumstances, strategic-to-operational planning focus may be required to perform a FID mission. Critical logistics requirements will need to be identified and planned. SF sustainment planners must consider ACSAs, theater Army and HNS relationships, transportation and ammunition requirements and constraints, identification of aerial ports of debarkation and seaports of debarkation, plus any other distribution infrastructure and associated capacity.

4-68. FID logistics civil support operations are limited by U.S. law. In earlier sections of this chapter, both financial management support to SF and legal support services were discussed. It is critical that SF units seek guidance from the CJA and RM prior to executing any type of sustainment support for the HN. Logistics support must be integrated into the overall theater FID plan, especially when the HN is involved in an active conflict. The following employment and sustainment considerations should be considered when providing logistics support as part of the SF FID effort:
- Develop definitive rules of engagement and protection measures for sustainment.
- Educate all members of the command on permissible activities in providing the logistics support mission.

- Build a logistics assessment file on logistics resources available in-country. Information such as local supply availability, warehousing and maintenance facilities, transportation assets, lines of communications, and labor force available should be included.
- Tailor the proper type of equipment maintenance and training sustainability packages to the needs of the HN.

LOGISTICS PLANNING FOR THEATER SECURITY COOPERATION PLAN EVENTS

4-69. TSCP events are tasked per fiscal year to SFG(A)s. TSCP planning follows four major milestones: the initial planning conference, the predeployment site survey, the mid-planning conference, and the final planning conference. Planning in support of TSCP events involves the SF battalion S-3, S-4, service detachment, and members of the participating SFODA. It is recommended that both SF elements and logisticians become familiar with applicable funding policy prior to TSCP planning and execution. MFP-11, NDAA Sections 1206 and 1208 Authorities, Equal Value Exchange, and ACSAs are examples of such policy. The TSCP initial planning conference is operationally focused. The TSOC will engage the supported country to glean training objectives for the duration of the exercise. This training concept will be shared with the SFG(A) for mission analysis. The battalion S-4 and service detachment logisticians will review the training objectives and determine initial logistical requirements.

4-70. The predeployment site survey follows the initial planning conference and is the unit's primary opportunity to execute the in-country logistics review. The unit will meet with the Embassy team, contracting officer, and TSOC logistics representative to review the standing logistics estimate and review in-country capabilities and constraints. An applicable logistics planning checklist is in Appendix A and a site survey checklist is located in Appendix D. The concept of support for TSCP events will vary depending on the supported country. Currently, much of the logistical support is facilitated by contract and operational funding. It is imperative that the deploying unit have a trained field ordering officer and pay agent. The type of TSCP training will further delineate which authorities will be used to finance the required training through the specific NDAA. Following the predeployment site survey, the unit will—

- Schedule strategic air requirements through the GSB transportation section.
- Forecast the required training ammunition.
- Coordinate the in-country ground movement plan.
- Coordinate home station airfield load-out and movement plan.
- Execute field ordering officer and pay agent training.
- Establish concept of support based on available HN assets, contracting requirements, and basic load.
- Process diplomatic clearances.

4-71. The mid-planning conference is used to further refine the training plan and concept of support. The TSOC and embassy country team personnel will provide critical information to assist in solidifying the operational and logistical support plans. The embassy country team will provide a list of approved vendors for use during the training period. Local vendors (contractors) for food, water, nontactical vehicles, minor construction materials, and so on are identified prior to the final planning conference. It is imperative that training units use the embassy-approved vendors to ensure quality products and reduce risk. The final planning conference solidifies the concept of support. The contracting officer is present and signs contracts for vendors that were selected during the mid-planning conference. Synchronization with the unit comptroller is critical while planning for TSCP events. The Department of the Army (DA) Form 3953 (Purchase Request and Commitment) must be sent to the theater contracting official to execute award of a contract.

4-72. A two- or three-person advanced echelon is typically used to move in advance of the main body to facilitate RSOI. The advanced echelon will develop the reception, staging, escort, and movement plan from the aerial port to the billeting and training area. The field ordering officer and pay agent are critical members of the advanced echelon party. Sustainment of the TSCP event is centered on properly executing contracts developed during the planning phase. Unforecast requirements are generally handled using OPFUNDs. The TSOC and embassy country team will assist with logistical issues during the training period. Any intratheater air movement performed during the training period is coordinated and executed by the TSOC mobility section.

4-73. At the completion of training, established rear echelon personnel will clear applicable in-country contracts. The TSOC (or supporting contracting officer) will typically arrive near the end of the training period to assist in the contract termination and payment, and OPFUNDs will be cleared by the issuing authority.

SUSTAINMENT SUPPORT FOR JOINT COMBINED EXCHANGE TRAINING

4-74. Upon receipt of a JCET mission or task, the SFODA should conduct a thorough mission analysis using the military decisionmaking process. As a result of this analysis, the team determines what is required to accomplish the assigned operational task. The SFODA then prepares the DSOR, which identifies those requirements the unit cannot satisfy with its organic assets or capabilities. The SFODA then staffs the DSOR. Once the DSOR is validated by the G-3/S-3 and sourced (funding, equipment, supplies, contracts, and so on), then it is just a matter of acquiring the goods and services. The command's fiscal year guidance will define the type of funding (overseas contingency operations versus baseline) that can be used for the required operation.

4-75. SF units are authorized to pay the incremental expenses associated with the training. An incremental expense means the reasonable and proper cost of rations, fuel, training, ammunition, and transportation for the HN. OPFUNDs play a role in the successful execution of a JCET. SFODA develops requirements for OPFUNDs during budget analysis and submits to higher for approval.

Chapter 5
Ranger Regiment

The Ranger Regiment HQ has an RSOD, and the battalions have organic RSCs to enhance the expeditionary capabilities of the regiment. With this structure, Rangers are organized with limited self-sustainment capability to support internal requirements for fuel, ammunition, maintenance, water production, and CUL for a short duration of time.

MISSION

5-1. The 75th Ranger Regiment's mission is to plan and conduct SO against strategic or operational targets in pursuit of national or theater of operations objectives. Rangers may conduct military operations independently or in concert with other SOF, Army conventional forces, and sister Services. These joint special military operations consist of deep penetration, direct action missions to capture or destroy critical enemy nodes and facilities, or recover designated personnel or equipment. These missions include conducting raids, seizing lodgments, and conducting noncombatant evacuation operations in permissive, uncertain, and hostile environments.

5-2. Rangers, unlike other SOF, are globally oriented rather than regionally oriented. Current force structure and contingency requirements preclude their apportionment to a specific GCC. They can deploy worldwide on short notice when a U.S. military presence would serve U.S. interests.

ORGANIZATION

5-3. The 75th Ranger Regiment (Figure 5-1) consists of a regimental headquarters and headquarters company (HHC), a Ranger Special Troops Battalion (Figure 5-2, page 5-2), and three Ranger battalions. In addition to mission command of three Ranger battalions, the regimental HQ may, if augmented, exercise operational control of Army conventional forces, logistics units, and other SOF for limited periods.

5-4. When required, the Ranger Regiment provides a liaison team with secure communications to the HQ of the supported unit commander that conducts operational and logistics coordination. The regiment has limited communication capability and is supported by ARSOF signal elements, as required.

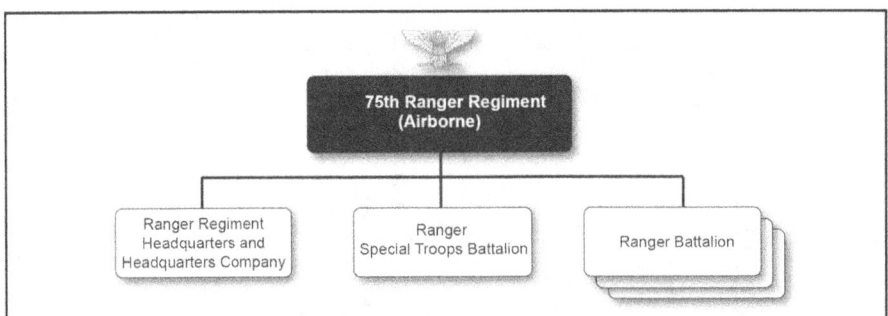

Figure 5-1. 75th Ranger Regiment (Airborne) organization

Chapter 5

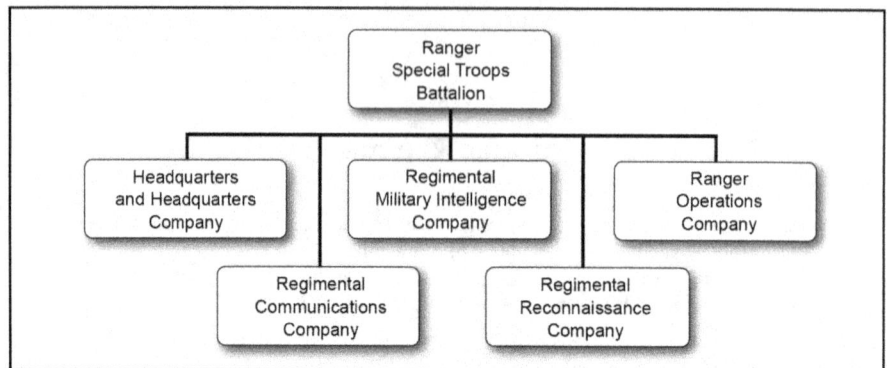

Figure 5-2. Ranger Special Troops Battalion organization

RANGER SUPPORT OPERATIONS DETACHMENT

5-5. The RSOD is assigned to the 75th Ranger Regimental HHC (Figure 5-3, page 5-3). The RSOD, under the direction of the SPO officer, provides centralized, integrated, and automated control for all logistics operations within the regiment. The RSOD provides information, input, or feedback to the battalion S-1 and S-4 for them to plan, coordinate, and provide the regimental commander an LCOP. It possesses the capability and expertise to integrate the regiment's logistics assets into the ASCC logistics structure.

MISSION

5-6. The RSOD coordinates with logistics operators in the fields of supply, ground and electronic maintenance, ammunition management, fuel, food service, aerial delivery, and movement management for the support of all units assigned or attached to the regiment. Its primary concern is customer support and increasing the responsiveness of support provided to subordinate units. The detachment continually monitors the support and advises the commander on the ability to support future tactical operations. The RSOD serves as the first point of contact for supported units' logistics requirements. The RSOD—

- Conducts continuous regimental-focused logistics preparation of the battlefield.
- Develops logistics synchronization matrices for regimental-level operations.
- Submits logistics forecasts to the supporting organization.
- Coordinates and provides technical supervision for the RSC's sustainment mission.
- Identifies tentative force structure and size to be supported.
- Coordinates the preparation of the SPO estimate on external support.
- Provides support posture and planning recommendations to the regimental commander.
- During regimental HQ deployment, sets up and supervises the logistics center (located in the tactical operations center) in conjunction with the S-4.
- Coordinates with the S-3 to determine air routes for supply.
- Provides centralized coordination for units providing support to the regiment.
- Analyzes contingency mission support requirements.
- Revises customer lists (as required by changing requirements, workloads, and priorities) for support of tactical operations.
- Coordinates external logistics provided by subordinate units.
- Advises the regimental commander on the supportability of missions and of shortfalls that may impact on mission accomplishment.

Ranger Regiment

- Serves as the single point of coordination for supported units to resolve logistics support issues.
- Plans and coordinates contingency support.
- Develops supply, service, maintenance, and transportation policies that include logistics synchronization.
- Plans and supports replenishment operations for all regimental units.

5-7. The SPO sergeant—
- Conducts continuous logistics preparation of the environment.
- Analyzes trends and forecasting requirements for supplies and equipment based on priorities and procedures.
- Coordinates major end-item resupply activities within the regiment.
- Coordinates activities internal to the SPO section.

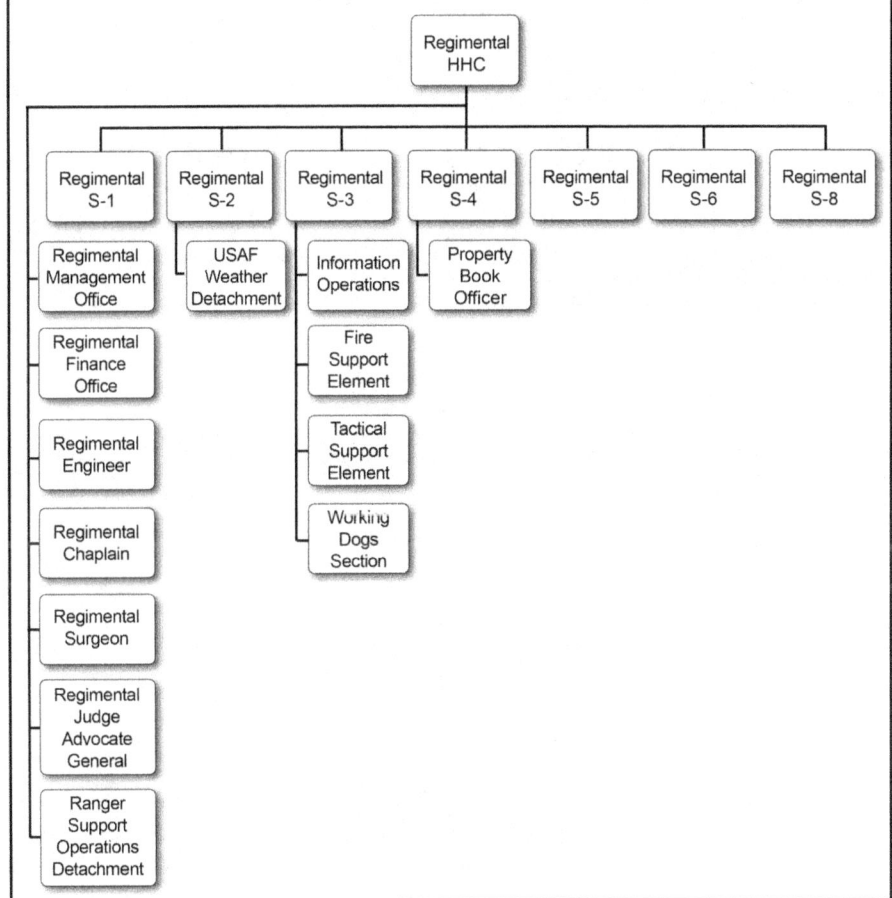

Figure 5-3. Ranger Regiment headquarters and headquarters company organization

Chapter 5

CAPABILITIES

5-8. The RSOD plans, coordinates, synchronizes, and integrates logistics for the regiment and subordinate battalions, including providing subject-matter expertise in quartermaster, transportation, and ordnance operations.

5-9. The RSOD facilitates support and sustainment planning for Ranger logistics operations, as required. The RSOD provides liaison and planning elements to ensure connectivity with theater of operations, HN, joint, and coalition logistical infrastructures. Liaison capabilities include identifying Ranger logistics and AHS requirements, conducting logistics support planning, coordinating for resources to satisfy requirements, and arranging access to CONUS wholesale points.

5-10. The RSOD coordinates daily logistics requirements, planning, and coordination for all external support requirements. It provides operational guidance to the regiment and maintains interface with the CONUS-based and theater of operations management functions. The RSOD coordinates with the JTF HQ, USASOC G 4, and national logistics agencies to ensure support and sustainment requirements are properly designated.

RANGER SUPPORT COMPANY

5-11. The RSCs are multifunctional logistics companies that are organic to each Ranger battalion (Figure 5-4) within the Ranger Regiment. They provide field maintenance; Class I, II, III (package) (bulk), IV, V, VII, and IX supply; water production with limited storage and distribution; transportation; aerial delivery; bare-base support; limited chemical, biological, radiological, and nuclear (CBRN) decontamination and reconnaissance; and food service. The RSC supports operations at the ISB and task-organizes necessary logistics assets to support Ranger units once they move from the ISB to mission support sites. Figure 5-5, page 5-5, depicts the RSC organization.

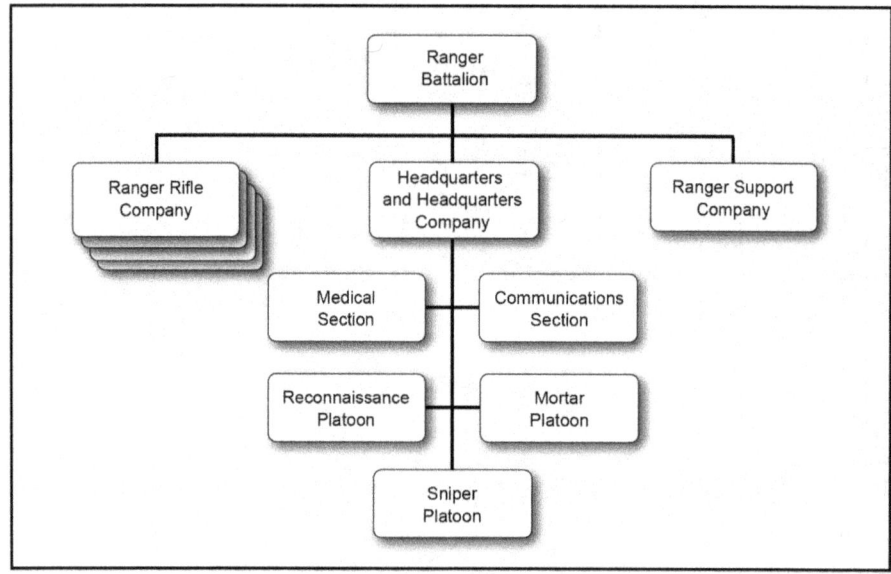

Figure 5-4. Ranger battalion organization

Ranger Regiment

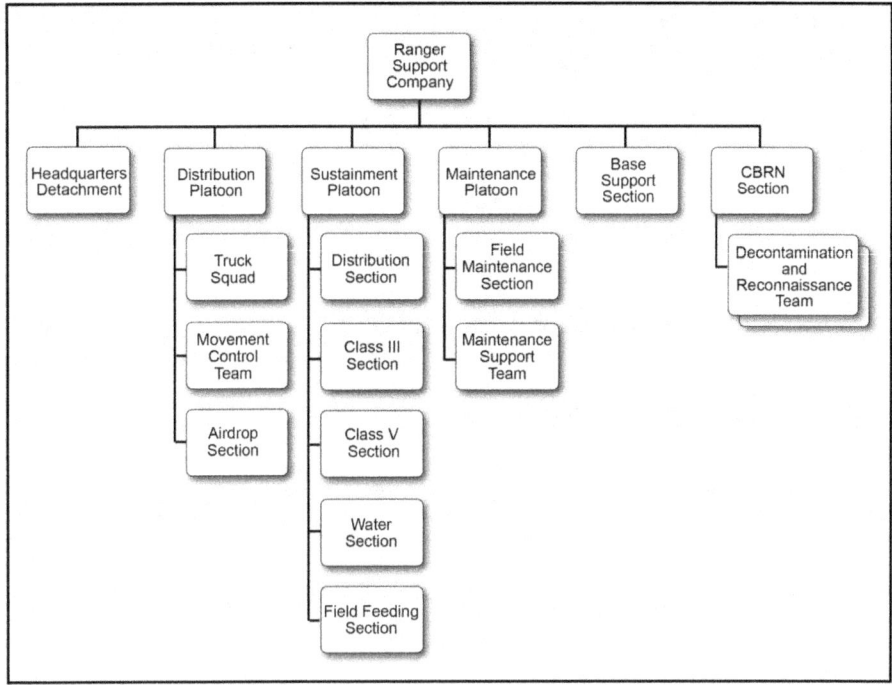

Figure 5-5. Ranger Support Company organization

MISSION

5-12. The RSC commander is the senior logistics provider at battalion level. He assists the battalion S-1 and S-4 with the logistics planning. He also provides information and feedback to the battalion S-1 and S-4 for their use in providing the battalion commander an LCOP.

5-13. The RSC is the primary CUL provider for all forces assigned or attached to the battalion. The RSC coordinates logistics requirements with the RSOD and JTF HQ. It can accept augmentation of, and employ, CUL assets from other Services and nations and integrate their capabilities into a cohesive plan that supports the operational concept. The RSC is capable, with augmentation and replenishment, of supporting all of the battalion's logistical requirements. When component forces are assigned to a SOTF, they will deploy with their organic support packages for Service-specific requirements and logistics support.

5-14. The RSC commander can execute the logistics plan IAW the battalion commander's guidance as developed by the battalion S-1 and S-4. The RSC commander responds directly to the battalion executive officer, who serves as the battalion logistics integrator and assists the battalion S-1 and S-4 in logistics synchronization and troubleshooting. His duties may require direct interface with the RSOD, joint and multinational forces, other SOF, and the TSC.

CAPABILITIES

5-15. The SPO functions are normally performed by the executive officer or one of the other company officers using METT-TC. Duties include the following:
- Provide continuous battle-tracking.
- Ensure accurate, timely tactical reports are received by the command post.

- In coordination with the battalion S-1/S-4, assist with developing the concept of support for the battalion OPORD.
- Conduct tactical and logistical coordination with higher, adjacent, and supported units, as appropriate.
- Oversee the development of the daily logistics packages.

LIMITATIONS

5-16. The augmentation of forces to a Ranger battalion may exceed the RSC's organic capability for sustainment. The RSC is not designed to provide all or part of the following logistics functions. An analysis of logistics shortfalls is prepared in order to coordinate for external support. Limitations of logistics functions include—

- Field services, to include—
    - Mortuary affairs—no planning, collection, processing, or evacuation without augmentation.
    - Shower, laundry, and clothing repair, which is not organic and support is provided by an external organization.
- Limited Class VIII/IX storage capability.
- Limited capability to reconfigure loads. Ammunition from echelons above the regiment must be in strategic or operational configured loads.
- Explosive ordnance disposal, which is provided by the maneuver enhancement brigade.
- Human resources support other than its own unit S-1 human resources operations. The RSC relies on external sustainment organizations to provide additional critical wartime personnel support.
- Legal support, which is limited to the assigned legal section.
- Optical fabrication and blood product management support.
- No organic aeromedical evacuation support.

HEADQUARTERS DETACHMENT

5-17. The RSC HQ detachment provides mission command, unit administration, internal supply support, billeting, discipline, security, training, and administration to assigned and attached personnel. It ensures that subordinate elements follow the policies and procedures prescribed by the battalion and RSC commanders. It directs the operations of its subordinate sections as well as the overall logistics operations, less medical, in support of the battalion.

DISTRIBUTION PLATOON

5-18. The distribution platoon provides transportation of materiel and personnel, movement control, and aerial delivery support functions.

TRUCK SQUAD

5-19. The truck squad provides the capability for transport of supplies and equipment. It also provides motor transport capable of moving containerized and noncontainerized cargo. Additionally, the truck squad provides aerial delivery support.

MOVEMENT CONTROL TEAM

5-20. The movement control team provides the management and coordination for movement control. It also coordinates the loading, off-loading, and transport of supplies, ammunition, explosives, equipment, materials handling equipment (MHE), and oversized equipment to and from aircraft or other transport, as directed.

AIRDROP SECTION

5-21. The airdrop section prepares general supplies and equipment for aerial resupply. It also provides personnel parachute-packing support.

SUSTAINMENT PLATOON

5-22. The sustainment platoon provides the battalion a single source for Class I (water), II, III (bulk) (package), IV, V, VI, and VII supply support to the battalion operations. The sustainment platoon receives, stores (limited), and issues Class II, III (package), IV, and IX supplies. It also receives and distributes, in coordination with the transportation platoon, Classes I and VI supplies from the field-ration issue point, and receives and issues Class VII supplies, as required. The platoon also maintains limited Class II, III (package), and IV supplies for the battalion. The ammunition transfer holding point section supports the battalion with Class V. The platoon HQ provides food service for assigned and attached units.

DISTRIBUTION SECTION

5-23. The distribution section consists of an SSA. This SSA uses SARSS-1 and related automated systems to provide ASL stock control, receipt, storage, and issue functions for both Army-common and SOF-peculiar items in garrison and deployed locations. The stock control supervisor must ensure that daily start-up and closeout procedures are followed IAW the schedule of operations established by the supporting HQ. Automated document processing and warehousing operations are conducted IAW AR 710-2, *Supply Policy Below the National Level*. The SSA—

- Operates the SARSS-1 system.
- Maintains a current ASL for all supported commodities.
- Processes receipts and requests for issues and turn-ins.
- Provides materiel release instructions to the warehouse section.
- Processes turn-ins to maintenance for repairable items.
- Performs periodic location surveys to ensure location accuracy.
- Processes inventory adjustments and creates necessary reports.
- Maintains coordination and provides general supervision over supporting signal assets.
- Establishes a storage and issue facility for all supported commodities.
- Performs receipt, storage, and issue of all supported commodities.
- Coordinates with SPO section for delivery and pickup of issued assets and turn-ins (to maintenance or disposal).
- Performs storage and inventory management activities as directed by stock control.

CLASS III SECTION

5-24. This petroleum, oils, and lubricants (POL) section provides the management, stockage, and delivery of all Class III (bulk) (package) items to the battalion.

CLASS V SECTION

5-25. This ammunition section manages the ammunitions and explosives training, basic load, and operational requirements. The section ensures that transportation of ammunition and explosives is achieved by current SOPs and Army and Air Force regulations. It is capable of operating one ammunition transfer point.

WATER SECTION

5-26. The water section provides potable water through the operation of one 125-gallons-per-hour reverse-osmosis lightweight water purifier operating 20 hours per day. The section provides a limited

Chapter 5

distribution using the Forward Area Water Point Supply System. The section is capable of storing a maximum of 9,000 gallons of water.

FIELD FEEDING SECTION

5-27. Class I is provided by the field feeding section. This section provides food service and food preparation for the battalion and organic and attached personnel in the SOTF. The RSC field-feeding section and the food service capability of supported units may merge to form a consolidated messing facility. This section distributes prepackaged and prepared food. It has the ability to prepare one heat-and-serve meal and one cook-prepared (A or B) meal per day. The section conducts remote feeding operations as required and maintains the unit's Class I basic load.

MAINTENANCE PLATOON

5-28. The maintenance platoon provides field-level maintenance on Army-common and SOF-peculiar automotive, ground-support, armament, construction, electronic/communications, quartermaster, and a wide variety of commercial equipment for the battalion and attached units. The platoon also maintains a Class IX shop stock and bench stock and provides recovery support.

FIELD MAINTENANCE SECTION

5-29. This section provides field-level maintenance on Army-common and SOF-peculiar automotive, electronics and communications, ground-support, armament, construction, quartermaster, and a wide variety of commercial equipment for the battalion and attached units, and provides reinforcing maintenance to the maintenance support team (MST).

MAINTENANCE SUPPORT TEAM

5-30. The maintenance section can field one MST that is organized to provide field maintenance for all vehicles organic to the battalion companies. The RSC commander sets the MST's priorities IAW the battalion commander's guidance. When deployed in support of a Ranger company, the MST operates under operational control of the company first sergeant and the maintenance noncommissioned officer in charge supervises the team. The scope and level of repair is based on METT-TC. The MST makes repairs as far forward as possible, returning the piece of equipment to the unit. During combat, the MST will perform battle damage assessment and repair, diagnostics, and on-system replacement of line-replaceable units. Emphasis is placed on troubleshooting, diagnosing malfunctions, and fixing the equipment by component replacement. If the tactical situation permits, the MST focuses on completing jobs on-site. The MST carries limited on-board combat spares to help facilitate repairs forward. If inoperable equipment is not repairable, due either to METT-TC or a lack of repair parts, the team uses recovery assets to assist the maneuver company and may recover inoperable equipment to the unit maintenance collection point or designated linkup point. The MST is fully integrated into the Ranger company's operational plans.

BASE SUPPORT SECTION

5-31. When augmented with troop labor, each engineer section supervises and provides the expertise to establish a forward support base. This section will provide limited protection support to SOTFs and will construct mission support sites, as required. The section provides engineer expertise to tie-in to D-Cell or Navy construction engineer battalion support for bare-base and warm-base construction, as required.

5-32. The base support section provides limited construction planning and execution capabilities. It maintains proficiency with common tool kits for carpentry, electrical, concrete, and masonry work. The section provides reachback capability to leverage technical expertise of government engineering assets (U.S. Army Corps of Engineers). The base support section is capable of—

- Limited vertical construction.
- Estimating a bill of materials and basic cost estimates.
- Basic concrete and masonry work.

- Site survey and basic site layout.
- Rough carpentry, framing, and finishing.
- Basic electrical work.
- Basic vertical and horizontal construction design.
- Repair and maintenance to existing structures.
- Construction management and contract oversight.
- Quality control and quality assurance.

CHEMICAL, BIOLOGICAL, RADIOLOGICAL, AND NUCLEAR SECTION

5-33. The CBRN section consists of two decontamination and reconnaissance teams. The teams are a battalion asset and are able to support CBRN operations at two locations simultaneously. The decontamination and reconnaissance teams can also make positive identification and perform decontamination of most known CBRN agents.

RANGER LOGISTICS SUPPORT

5-34. The Ranger Regiment is an austere organization with organic logistics capability that relies on support from home station or prepackaged supplies during the initial stages of the deployment. As the theater of operations matures, replenishment is provided by the TSC or joint logistics providers within the JOA. The regiment primarily receives logistics support from the RSOD and the Ranger battalion's RSC.

5-35. The Ranger force is limited by airframes for transport of both personnel and equipment. Whenever possible, the contingency stocks are augmented from theater of operations sources to reduce the number of aircraft required deploying and supporting a Ranger force.

5-36. Rangers will deploy in support of an OPLAN or CONPLAN. Therefore, the logistics concept of support must be flexible and be tailored to support the operational requirement. As a member of USSOCOM, Rangers receive support from installations under 10 USC, and either the SOTF or the land component command, depending upon the task organization.

5-37. The Ranger Regiment requires external air and ground transportation for deployment and most infiltrations. This resupply system allows the regiment to deploy rapidly and be self-sustaining until the RSOD can coordinate with the SOTF and ALE to obtain support from the TSC or joint logistics providers within the JOA. This system also allows deploying Rangers to take what supplies they need or the airflow will allow, and enables follow-on aircraft to build up required supplies quickly.

ANNISTON ARMY DEPOT CONTINGENCY STOCKS

5-38. Anniston Army Depot maintains prepackaged contingency stocks of ammunition and Class VIII items for the 75th Ranger Regiment. These contingency stocks are pallets which are easily transferred to a departure airfield by ground vehicle and then transported by C-17 aircraft. This unique capability provides the regimental commander flexibility above and beyond support available in the theater of operations, as well as being able to function in austere or undeveloped theater of operations.

ARMY HEALTH SYSTEM SUPPORT

5-39. The Ranger Regiment has a medical section for which the regiment's surgeon has supervisory oversight. The surgeon's oversight includes responsibility for all AHS training opportunities in the regiment. The regiment's medical section provides AHS support for the regiment's HHC and Ranger Special Troops Battalion. It also plans and coordinates theater of operations AHS support for the ISB and SOTF operations, and medical support at the Ranger objective. This support encompasses AHS augmentation; ground, rotary-wing, and fixed-wing evacuation; and Class VIII resupply. Additionally, each Ranger battalion has a medical section with a surgeon who has supervisory oversight. Both regiment and battalion medical staffs have experience planning and leading joint casualty collection points.

5-40. The Ranger AHS mission is to provide combat trauma management to treat the wounded and injured and to save lives; to plan and conduct MEDEVAC and casualty evacuation (CASEVAC) for Ranger operations; to conduct a daily sick call; to plan, conduct, and instruct AHS training for individual Rangers and medical personnel; and to manage AHS administrative duties for all assigned personnel.

TACTICAL MEDICAL EVACUATION

5-41. Ranger forces have limited CASEVAC assets, and must rely on the SOTF, TSC, and ASCC aviation support for air MEDEVAC. The only means of tactical evacuation Ranger forces have in the target area are a limited number of medical SOF vehicles. Ranger medics have a habitual training relationship with other SOF units that have some ground MEDEVAC platforms that augment the Ranger capability on a regular basis.

5-42. Generally, wounded or injured Rangers are moved in the local target area by buddy-carry or by medical SO vehicles to a casualty collection point where triage and trauma management occurs. The casualty collection point is normally located near a helicopter landing zone or fixed-wing aircraft parking area on a target airfield. The wounded are loaded onto air assets at or near the target for evacuation to the ISB or other trauma facility.

RELIGIOUS SUPPORT

5-43. The Ranger regimental HQ has a UMT for which the regimental chaplain provides supervisory oversight. The regimental UMT provides religious support and pastoral care for the regimental HQ. Each Ranger battalion also possesses—as an organic asset—a battalion UMT. The regimental chaplain provides direction and supervision for the collective Ranger regimental UMT.

5-44. Ranger UMTs provide worship services, religious rites, sacraments, ordinances, pastoral care, counseling, and crisis and emergency ministry to assigned Rangers and their family members, as well as to authorized civilian employees. To extend their ministries during decentralized operations, Ranger UMTs may develop a support network of Rangers who serve as unit religious coordinators at company and platoon levels.

Chapter 6

Army Special Operations Aviation Command

The ARSOAC supports other SOF units by planning and conducting special air operations in all operational environments. Its specially organized, trained, and equipped aviation units give the joint force special operations component commander the capability to infiltrate, resupply, and exfiltrate SOF elements engaged in all types of missions and environments.

MISSION

6-1. The ARSOAC headquartered at Fort Bragg, North Carolina, conducts and supports special air operations by clandestinely penetrating hostile and denied airspace. ARSOAC units can operate in harsh environments and across the range of military operations. They also support SOF in conducting joint, combined, interagency, liaison, and coordination activities in support of the USSOCOM commander and the GCC's concept of operations. The participation of the SOAR in the ARSOF core activities varies based upon the type of conflict, the environment, and the scope of the operation.

ORGANIZATION

6-2. The ARSOAC organizes, mans, trains, resources, and equips Army SOA units to provide responsive SOA support to SO. Additionally, the command serves as the USASOC aviation staff proponent, and includes a technology applications program office, a flight detachment, a systems integration management office, a regimental organizational applications element, a SOA training battalion, and the 160th SOAR(A) (Figure 6-1, page 6-2).

6-3. The 160th SOAR(A) consists of an HHC and four SOA battalions. The units can conduct and support SO missions for the ARSOF commander or for the TSOC. The 160th SOAR(A) can be task-organized based on expected missions, the requirements of the units being supported, the environmental conditions in the theater of operations, and sustainment requirements. The unit can task-organize missions around one of the SOA battalions. With proper personnel and equipment augmentation, the 160th SOAR(A) battalion commander and his staff could also serve as a joint special operations air component commander. When two or more battalions are required in the theater of operations, the regimental commander could serve as the joint special operations air component commander.

6-4. The 160th SOAR(A) (Figure 6-2, page 6-2) supports other SOF units by conducting special air operations in all operational environments. The specially organized, trained, and equipped aviation units give the joint force special operations component commander the capability to infiltrate, resupply, and exfiltrate SOF elements engaged in all core activities, missions, and environments. FM 3-76, *Special Operations Aviation*, provides additional information on SOA units.

Chapter 6

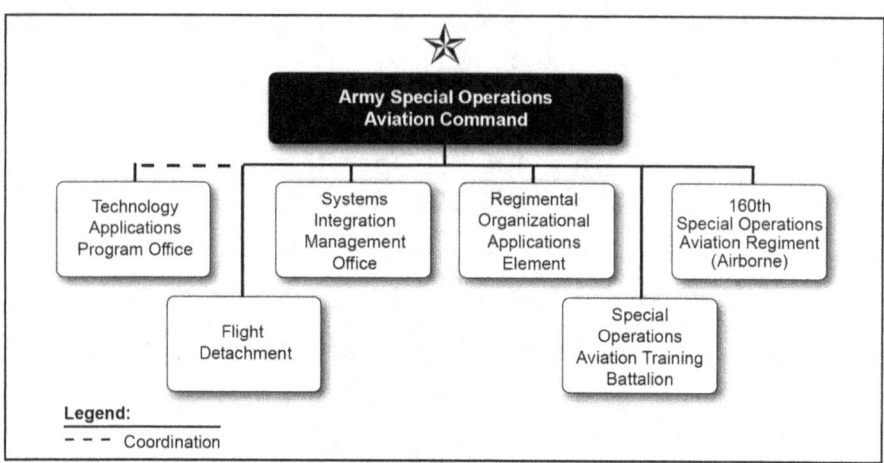

Figure 6-1. Army Special Operations Aviation Command

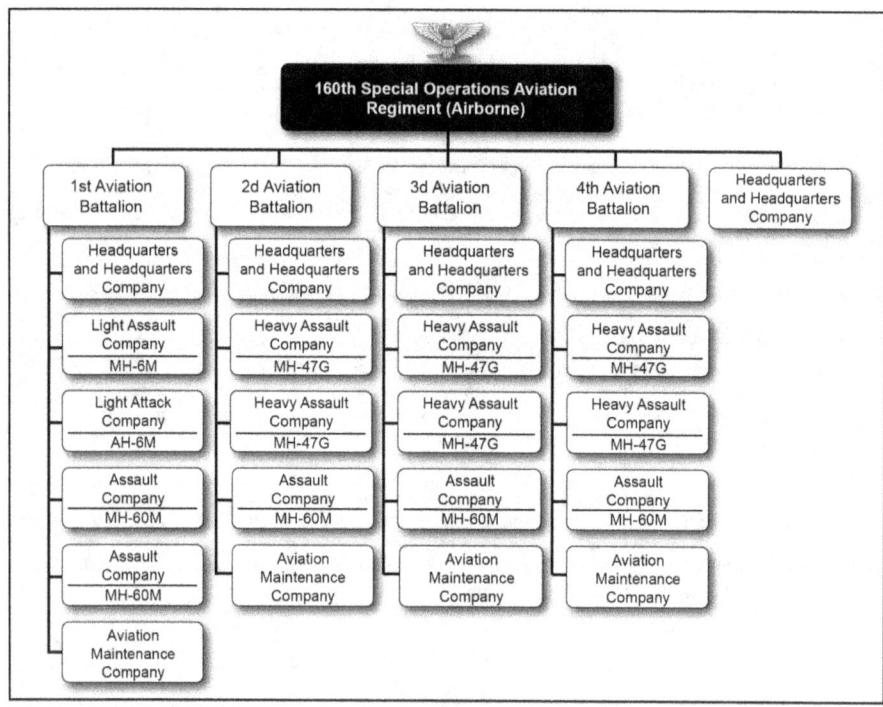

Figure 6-2. 160th Special Operations Aviation Regiment (Airborne) organization

PLANNING CONSIDERATIONS

6-5. The collocation of SOA assets with other SOF reduces distribution problems and facilitates use of the airlift. SOA units also have limitations that planners must consider. They have—

- Limited organic logistics support capability.
- No food service support or water storage capability. For extended deployments, food service support is required to support 24-hour operations and varied aircrew schedules.
- Insufficient manpower to provide adequate airfield security; augmentation is required.
- No organic billeting for personnel. Extended deployments require facilities to billet aircrews. The facilities should be climate-controlled and compartmented to support varied complicated aircrew rest schedules.
- Insufficient organic transportation to conduct their own unit movement. Augmentation is required from the supported unit or theater of operations transportation assets.
- The requirement for stovepipe requisition and distribution systems to resupply SOA-peculiar Class II, V, and IX items.
- No ground support assets necessary to accept supply point distribution. The unit distribution method of resupply is required.
- No ability to effect integration into the airspace control system. SOA units require support or augmentation for airspace deconfliction and tactical air support coordination.
- A small-sized force and the inability to regenerate personnel or equipment rapidly.
- Limited Class III bulk capability. SOA units rely heavily upon theater of operations support.

CONTINGENCY PLANNING

6-6. In contingency planning, ARSOF units, in conjunction with the 528th SB(SO)(A), prepare a support plan. The support plan identifies support requirements for OPLANs and CONPLANs in a bare-based SOR, down to the user level. The ASCC coordinates with theater of operations support organizations to fulfill requirements and prepares a support plan identifying support relationships and shortfalls.

6-7. All logistics operations constantly strive to maintain units at a desired level or resource. To maintain the desired level, planners must—

- Maximize the use of existing fixed facilities.
- Limit logistics requirements to mission essentials and acceptable risk.
- Minimize the handling of supplies.
- Concentrate maintenance on returning major end items to service.
- Rely on air lines of communications for rapid resupply.
- Anticipate high attrition of supplies while performing missions in denied areas.
- Identify to the ASCC as early as possible those items that require operational floats or other special logistics arrangements.
- Make maximum use of HNS, including local and third-country resources.
- Conduct threat assessment.
- Conduct risk assessment.

CRISIS ACTION PLANNING

6-8. Crisis action planning is based on current events and is conducted during time-sensitive situations and emergencies using assigned, attached, or allocated forces and resources. Planners for crisis action planning base their approach on the actual circumstances that exist at the time planning occurs. They follow prescribed procedures that parallel contingency planning, but are more flexible and responsive to changing events and time constraints.

6-9. ARSOF logistics planning must take into consideration the bare-based requirements to support operations. In the early stages of any deployment, ARSOF will normally be required to establish separate ISBs and eventually expand the number of support bases to meet mission requirements. To maintain the desired level of support, meet operational tempo projections, and provide flexibility, each planner must be able to meet current requirements and to simultaneously plan for future SO requirements. Planners must first consider the existing infrastructure in the theater of operations. Using this infrastructure as a baseline, planners then integrate, consolidate, and cross-level resources to maximize logistics support.

LOGISTICS SUPPORT

6-10. Conventional logistics organizations and procedures are normally adequate for SOA requirements. Standard procedures are in place to handle the few SOF-unique requirements through the TSOC and the ALE. The TSC provides RSOI, follow-on support, and sustainment of theater of operations Army forces. SOA has some key differences that impact the type of support required for RSOI and sustainment. The following conditions occur often enough that they must receive special consideration during logistics planning:

- Forward-deployed units are usually in isolated and austere locations. Distribution is an essential consideration.
- Some special equipment exists; however, most equipment is Army-common, and organic assets can maintain it.

6-11. When SOA is normally attached to a SOTF with SF or Ranger logistical organizations, the SF GSB or the Ranger regimental support company will be the CUL provider. It will arrange support by coordinating requirements through the 528th SB(SO)(A), ALE, TSOC, and TSC, and by reachback through the 528th SB(SO)(A) to USASOC and USSOCOM for SOF-unique support. The SF or Ranger logistics organizations are joint- and multinational-capable in that they can accept augmentation of, and employ, CUL assets from ARSOF and other Services and nations. They will then integrate their capabilities into a cohesive plan supporting the commander's operational concept. When ARSOF are assigned to a JSOTF, they will provide their organic support packages for Service-specific and common logistics support.

6-12. The TSOC, ALE, SO liaison element, and logisticians coordinate with the TSC to develop plans and subsequent orders. They assist in the development of, and implement, directives the commander issues to support the SOA assigned to the GCC. The TSOC and ALE advise the GCC on the appropriate command and support relationships for each SOA mission. The ALE keeps the 528th SB(SO)(A) informed of the status of theater of operations supporting plans and logistical shortfalls.

6-13. The GCC supports the SOA in his AOR. The regiment's logistics planners, with the assistance of the ALE, identify support requirements in the planning phase. They must also identify the logistics shortfalls for inclusion in the GCC's risk assessment. If the TSC cannot support the SOA, it must raise the shortfall to the supported GCC for resolution.

SPECIAL OPERATIONS AVIATION LOGISTICS REQUIREMENTS

6-14. SOA units require certain logistics support for which they have no organic capability. The SOA services requirements are as follows:

- Logistics planners must identify and procure tentage for the task force operating in an austere environment. When available, fixed, climate-controlled billeting is optimal for flight management. SOA must identify billeting requirements.
- SOA has limited airdrop resupply and equipment maintenance capability. However, it can provide airborne insertion of a FARP and mission command elements. SOA must identify follow-on airborne requirements. Coordination is through the TSC or JTF.
- The regimental S-4 oversees deploying aviation life support system personnel and equipment based on mission variables, mission profile, and duration of the mission. The aviation life support officer ensures pre-inspections of personal equipment, protective armor, climate kits, and mission-specific equipment. Aviation life support system specialists deploy with limited

backup equipment to support the deploying task force. Additionally, the aviation life support system section provides search and rescue swimmers for overwater operations.
- The SOA force requests mortuary services, as required. Requests are coordinated through the SOTF and the TSC.
- SOA has no food-service capability. Because of mission duration and times, the task force requires rations during 24-hour operations. SOA must rely on the supported unit to provide food service. The SOR must identify food-service requirements.
- Based on duration of the operation, the SOA force may require laundry and shower services. When developing the SOR, logistics planners must include water requirements for these services into the total water requirements.
- SOA has additional water requirements to wash aircraft and flush engines to prevent corrosion during operations in austere environments. Logistics planners must compute these water requirements and identify them in the SOR. Table 6-1 lists the minimum water planning requirements for each type of organic aircraft in remote operations. This water requirement is for manual washing of aircraft and engine flushing on a daily basis. In desert environments, increased water requirements may be required because of the effects of fine sand on the aircraft.

Table 6-1. Water requirements for aircraft washing and engine flushing (gallons)

Type of Aircraft	Water Requirements (Gallons)		
	Fuselage	Engine	Total
MH/AH-6	25	10	35
MH-60	30	20	50
MH-47	50	21	71

6-15. SOA collocation of assets with other ARSOF or conventional units reduces external logistics and FHP distribution problems, and facilitates use of regimental airlift. The following describe organic support capabilities and limitations when not collocated with logistics and sustainment organizations:
- The force deploys with a basic load of meals, ready to eat, for initial sustainment. It has no organic food-service or water storage capability.
- The force deploys with a basic load of Class II supplies for initial sustainment. It has limited document management resources, such as computers, copiers, and shredders.
- Theater of operations pipeline sustainment, joint assets, or in-country sources provide bulk fuel in the theater of operations. During sustained operations, heavy expanded mobility tactical truck fuelers deploy if airlift or sealift is available from the TSC to provide fuel support at the ISB or forward staging base. Then the fuelers can establish limited FARPs. SOA units can deploy the equipment by airborne or air landing methods to establish 500-gallon-blivet or 20,000-gallon-bladder FARPs, usually in support of a tactical operation. SOA units do not have the capability to conduct long-term sustainment operations without bulk resupply from theater of operations assets.
- Theater of operations assets must deliver bulk resupply, as the regiment does not have the organic capability to transport large quantities of fuel. Because of the high operational tempo of the unit, the fuel requirement is higher than it is for a similarly sized conventional force. The regiment deploys with a basic load of Class III packaged POL for initial sustainment. When appropriate, SOA requires aerial refueling support for long-range missions. The SOA unit must identify the refueling requirements as soon as possible.
- Identification of Class IV materiel occurs based on mission requirements in the SOR. Because of the limited space on USAF Reserve intertheater airlift allocated for deployment, coordination must occur for pre-positioning and HNS.
- The SOA unit deploys with a basic load of common and specific Class V supplies. Planners schedule airlift and configure resupply and follow-on ammunition packages for delivery based on the mission. The TSC or JTF coordinates ammunition resupply from available sources in the

Chapter 6

theater of operations. The SOA unit logistics planners identify common Class V requirements using the SOR. The SOA unit has a limited capability to transport or store large quantities of Class V supplies and relies on theater of operations transportation and storage.

- Units deploy with Class VI items for initial sustainment (usually a 14- to 30-day supply), when available. Health comfort packets arrive in the theater of operations upon establishment of the logistics system. The SOA unit controls weapons systems and replacement aircraft from base stations using limited operational readiness floats. The deployed force requests airframes, weapons systems, and aviation parts through the J-4 to CONUS logistics channels. The units' S-4s coordinate with appropriate activities and item managers for immediate release of replacement systems.
- The SOA unit's flight surgeons develop their deployment load of Class VIII supplies based on the mission variables of METT-TC for initial sustainment. The force then integrates into the joint or theater of operations HSS system for resupply and sustainment.
- Forward support packages deploy with the force. These packages include Class IX air and armament parts and contractor logistics items. The unit's S-4 directs the deployment of the forward support packages based on mission variables and availability of air lines of communication for initial sustainment and follow-on resupply. If air lines of communication are unavailable after deployment for a brief period of time, the forward support section coordinates with the regiment aviation maintenance office and directs additional items to accompany the standard forward support package.
- The ALE coordinates, through the TSC or 528th SB(SO)(A), Class X supplies for CMO, based on mission variables. Coordination occurs with the JTF or JSOTF staff.

SUSTAINMENT FOR DEVELOPED AND UNDEVELOPED THEATER OF OPERATIONS

6-16. Once the SOA unit is integrated in the ASCC logistics system, the regiment will coordinate with the supported theater of operations for logistics resupply. The supported ASCC has an ARSOF ALE charged to coordinate logistics for operating in the theater of operations. The ALE is a key element in ensuring logistical requirements meet the SOA unit's requirements.

6-17. The ASCC receives the validated SOR. The TSC reviews documents (usually during initial and in-progress planning conferences) with the units to determine availability of support and services. The ALE planners coordinate key elements in the theater of operations logistics structure, particularly the TSC, to support SO. The essential element of support is the establishment of scheduled intertheater and intratheater airlift. Coordination of movement from the home station to the theater of operations is through the 528th SB(SO)(A) to the USASOC G-3.

6-18. Coordination of movement within the theater of operations is through the TSOC or the JTF joint movement center, with approval for airlift use coming from the GCC. This transportation support is the hub of logistics support, since many SOF-unique repair parts, test sets, and associated tools are unavailable in normal theater of operations supply systems. This airlift transports SOF-unique items from origin to the aerial port of debarkation. If the port of debarkation is the destination airfield in the supported theater of operations, the unit (if within range) picks up the repair parts or scheduled intratheater transportation delivers the parts to the destination airfield.

6-19. The deploying SOA unit must accomplish the following logistical tasks:
- Develop the SORs based on OPLANs and mission plans.
- Submit the SORs through operational channels for validation by the TSOC as early as possible but not later than the suspense date.
- Deploy with sufficient required basic loads.
- Schedule additional supplies on later flights as priorities allow.
- Resource the following key personnel to facilitate parts and equipment collection and transfer:
  - Ensure key personnel coordinate with the forward support package manager (deployed) and the SSA at Fort Campbell, Kentucky.
  - Coordinate with battalion S-4 representative and production control forward (deployed).

- Coordinate with battalion S-4 and production control rear, who have access to unit technical supply sections able to conduct lateral searches for required items needed forward.
- Provide regiment S-4 representative in the regiment emergency operations center an information copy of requests (message traffic, fax transmissions) from deployed assets or units upon receipt.

6-20. Upon receipt of a mission or the notification of an impending mission, the SOA HQ begins planning the operation or contingency. Upon notification of authorization to deploy forces, the SOA HQ—

- Implements a 24-hour emergency operations center.
- Provides a forward support package manager for the deploying task force.
- Reviews with the regiment S-3 the SORs from the deploying force and submits these requirements to 528th SB(SO)(A), JSOTF J-4, and the TSC.
- Provides 24-hour oversight of activities of the SSA, aviation life support system, PBO, organizational clothing and individual equipment, and regimental aviation maintenance officer for aviation-intensive-managed items release.
- Provides PBO or materiel management for deployed assets.
- Coordinates directly with the designated direct support unit, under "direct liaison authorized."
- Provides a deployment DOD activity address code to the deploying task forces.
- Coordinates (through the ALE) for all local purchases of items not readily available from the Army supply system and SO sources of supply.

HEALTH SERVICE SUPPORT/FORCE HEALTH PROTECTION

6-21. The SOA is assigned a flight surgeon, a clinical psychologist, and several special operations combat medics (SOCMs) who are flight-medic-qualified. The regiment does not have physician assistants assigned to its organization. The regiment is dependent upon the theater of operations HSS/FHP assets for Role 2 and above support. Table 6-2 lists the HSS/FHP assets.

Table 6-2. Medical personnel authorizations for the special operations aviation regiment

Unit	Personnel
Special Operations Aviation Regiment	Group Surgeon, Area of Concentration 61N, Major (MAJ)
	Clinical Psychologist, Area of Concentration 73R, MAJ
	Senior Flight Medical NCO, MOS 01EW1, SOCM, Sergeant First Class (SFC)
	Flight Medical NCO, MOS 91EW1, SOCM, Sergeant (SGT)
	Flight Medical Specialist, MOS 91EW1, SOCM, Specialist (SPC) (2)
Special Operations Aviation Battalion	Battalion Surgeon, Area of Concentration 62N, Captain (CPT)
Additional Assets	Approximately 15 Flight Medical Specialists, SPC, Distributed Throughout the SOAR
NOTE: Flight medical sergeants and specialists have the special qualifications identifier (SQI) designating flight status.	

FUNDING AND FINANCE SUPPORT

6-22. The finance battalion in the theater of operations provides support for funding and finance service, or the finance battalion supporting the SOA unit in garrison (as determined by the task force commander) may provide the service. Funding and finance support includes—

- Providing funds to the ALE or other agents.
- Coordinating resupply of funds, materiel, and services in the theater of operations.

Chapter 6

- Coordinating currency exchange with the appropriate embassy or agency.
- Paying local vendors and contracts.

ENGINEER SUPPORT

6-23. The SOA unit is dependent upon ASCC and theater of operations engineer units for support and sustainment. When available, engineer units conduct a variety of missions, to include the following:

- Engineer reconnaissance teams may assist in reconnaissance missions to locate possible sites for FARPs, landing zones, or advanced operating bases.
- Engineers provide current mine threat overlays that may impact ground operations. They clear obstacles and possible booby traps.
- Engineers support countermobility by providing hasty protective row minefield training and by installing obstacles to disrupt, turn, fix, and block enemy forces.
- Engineers construct berms and trenches to protect holding areas and FARPs. They help construct wire obstacles around the perimeter. They also help in training camouflage techniques.
- Engineers perform tasks to ensure the continuous sustainment for forward-deployed assets, to include replacement of tactical bridges, support facilities, and area damage control. Tasks also include constructing, maintaining, and repairing combat roads and trails, main supply routes, and lines of communication.
- Engineers provide terrain data in support of air and ground operations. Terrain data help identify possible air corridors, FARP operations, potential landing zones, pickup zones, and terrain that can mask movement.

FORWARD ARMING AND REFUELING POINT OPERATIONS

6-24. The 160th SOAR(A) has an organic airborne forward arming and refueling section that provides Class III (B) and (V) support for operational units. The airborne forward arming and refueling section can rig for airdrop and operating 12-, 16-, or 32-foot Type V platforms with FARP equipment. FARP personnel can operate MH/AH-6, defensive armed penetrator (DAP), MH-60, and MH 47 FARPs. They can also operate during joint fixed-wing refueling operations in forward areas. Due to the high volume of fuel required for the MH-47 and MH-60, the tactical airdrop FARPs are usually used in support of the MH/AH-6. This aircraft has limited range and lack of in-flight refueling capabilities.

LOGISTICS IN DEVELOPED AND UNDEVELOPED THEATER OF OPERATIONS

6-25. The uncertainty of today's world presents great challenges for supporting and sustaining ARSOF. These challenges include drug trafficking, natural and man-made disasters, regional conflicts, insurgencies, and terrorist and conventional threats with state-of-the-art weapons. A challenge for ARSOF logistics emerges from the small size of these operations. Small-scale operations will result in smaller, less-developed theater of operations with little to no dedicated ASCC logistics support structure.

6-26. In the early stages of an operation or during crisis response and limited contingency operations, the 528th SB(SO)(A)—in addition to performing RSOI—may be responsible for establishing the theater-level stockage base and providing logistics support to units deployed forward into their AOs. As the theater of operations grows and matures, this sustainment function will transition on order to the sustainment brigade tasked to provide theater of operations distribution and/or to an operational-level sustainment brigade in the theater of operations.

DEVELOPED THEATER OF OPERATIONS

6-27. In a developed or mature theater of operations, a sustainment base sets up within the theater of operations. Pre-positioned war reserve materiel stock and operational project stocks are in place, and foreign nation support agreements exist. TSC capabilities are normally sufficient to support and sustain

ARSOF. In cases where the TSC is unable to fill ARSOF logistics requests or requirements, the 528th SB(SO)(A) or ALE will exercise their reachback capability to USSOCOM or USASOC to fulfill the requirement.

UNDEVELOPED THEATER OF OPERATIONS

6-28. An undeveloped theater of operations does not have a significant U.S. theater of operations sustainment base. Pre-positioned war reserve materiel stock, theater of operations operational project stocks, and foreign nation support agreements are minimal or nonexistent. When an ARSOF unit deploys into an undeveloped theater of operations, it must bring sufficient resources to survive and operate until the ASCC establishes a bare-based support system or makes arrangements for HN and third-country support. The bare-based support system may function from CONUS, afloat (amphibious shipping or mobile sea bases), or at a third-country support base. The bare-based support system relies heavily on intertheater airlift or sealift for resupply.

CONTRACTING

6-29. ATTP 4-10 and JP 4-0 explain the procedures for obtaining contracting support. Depending on the operational situation and its associated risks, a variety of support functions exist on the battlefield that a contractor can provide or augment. All functions other than those inherently governmental in nature (defined as armed combat, command and control of U.S. military and civilian personnel, and government contracting) or functions covered by HNS agreements, may be suitable for contractor support.

6-30. Contracting support on the battlefield is an integral part of the overall process to obtain supplies, services, and construction in support of SOAR operations, and is a critical capability required in an underdeveloped theater of operations. Contracting support can augment existing capabilities, provide expanded sources of supplies and services, bridge gaps in the deployed force structure, leverage assets, and reduce dependence on U.S.-based logistics. Contracting for supplies and services lessens the requirements normally performed by logistics personnel. Contracting personnel should arrive with, or before, the lead ground elements to establish contracting operations. They should depart with, or after, the last ground element to close out the operations. Contracting personnel should establish their operations with, or near, local vendor bases to support deployed forces. Commanders should understand that the contracting officer may need to reside and operate outside a SOTF to be accessible to the local vendor base. Additional information on contracting support is in Chapter 11 of this publication.

This page intentionally left blank.

Chapter 7
Civil Affairs Brigade

CA provides the military commander with expertise on the civil component of the operational environment. The commander uses CA capabilities to analyze and affect the populace through specific processes and dedicated resources and personnel. As part of the commander's CMO, CA conducts operations nested within the overall mission and intent. CA significantly helps ensure the legitimacy and credibility of the mission by advising on how to best meet the moral and legal obligations to the people affected by military operations. The key to understanding the role of CA is recognizing the importance of leveraging each relationship between the command and every individual, group, and organization in the operational environment to achieve a desired effect.

MISSION

7-1. The 95th CA Brigade (A) mission is to rapidly deploy regionally focused initial-entry CA planning teams, CMO centers, CA battalions, and companies to plan, enable, shape, manage, and execute CAO in support of a JSOTF, TSOC, joint force special operations component, or interagency organization. The brigade supports the commander by engaging the civil component of the operational environment to achieve CMO or other stated U.S. objectives and ensure the sustained legitimacy of the mission and the transparency and credibility of the military force before, during, or after other military operations.

ORGANIZATION

7-2. The 95th CA Brigade (A) consists of a brigade HHC and five regionally aligned battalions. The battalions consist of five companies composed of five CA teams, each designed to support the brigade's worldwide mission (Figure 7-1, page 7-2). More detailed information is in FM 3-57, *Civil Affairs Operations*.

LOGISTICS SUPPORT

7-3. The type of operation, deployment sequence, unit-basing, and AOR shape the logistics environment for CA forces. Geographic TSC organizations and procedures are normally adequate for CA requirements. 528th SB(SO)(A) and ALE procedures are in place to handle the few CA-peculiar equipment requirements. The ASCC, assisted by the ALE, provides RSOI and follow-on support and sustainment of ASCC forces, including ARSOF. The following conditions occur often enough that CA units must receive special consideration during logistics planning:
- Forward-deployed CA units are usually in isolated, austere locations. In such cases, distribution of the support requirement is the key consideration.
- Although a requirement may exist for some special equipment, most equipment is Army common and organic logistics units can maintain the equipment.

Chapter 7

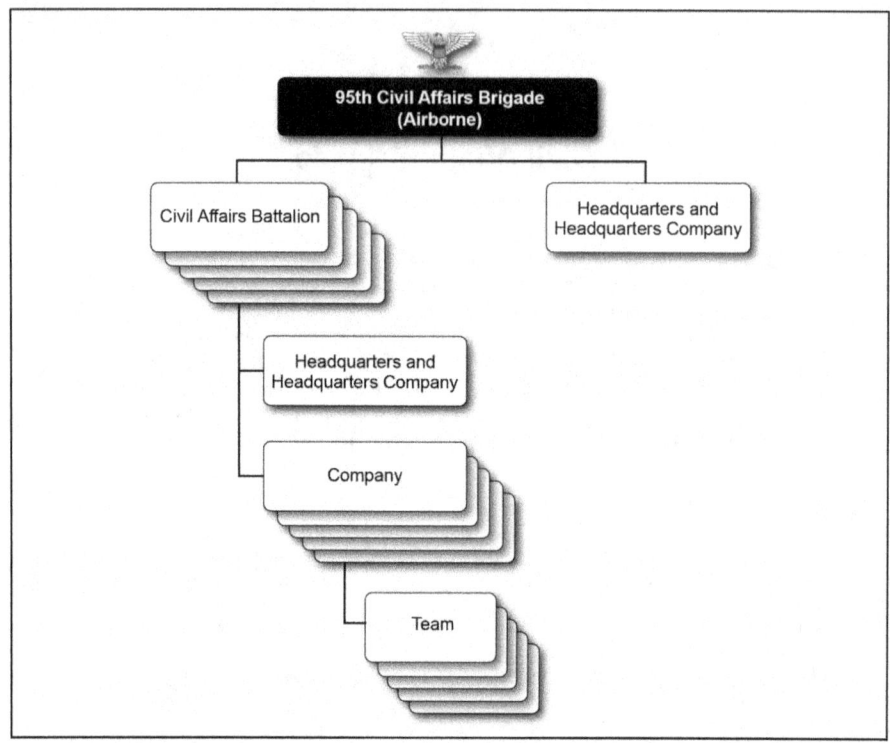

Figure 7-1. Active Army Civil Affairs (Airborne) organization

7-4. The GCC through the TSOC tasks missions to CA forces. The GCC's staff works closely with the TSOC, ALE, and the TSC to articulate the CA requirements. The GCC establishes priorities and allocates the available resources to accomplish each mission. The GCC develops the theater of operations support plan of theater of operations logistics organizations that include CA logistical requirements.

7-5. The TSOC establishes the command relationship involving CA forces within the theater of operations. CA logistics planners coordinate with the 528th SB(SO)(A) and ALE to develop plans and subsequent orders or to implement directives. These requirements are coordinated with the TSC and integrated into the overall logistics support plan. The ALE keeps the 528th SB(SO)(A) informed of the status of GCCs' supporting plans and projected CA logistic shortages.

7-6. CA planners identify the logistics support requirements in the planning phase. The planners must also identify the logistics shortfalls for inclusion in the GCC's risk assessment.

7-7. CA units should develop a concept of support and logistics estimates during the military decisionmaking process. The CA units should communicate their sustainment requirements early on with the 528th SB(SO)(A) home station SPTCEN and the ASPO cell. Doing so enables the theater of operations ALEs to assist in planning and integrating the CA units' sustainment requirements into the theater of operations sustainment infrastructure. When a CA unit is attached to an ARSOF-led JSOTF with organic ARSOF logistics units (GSB/RSC), the ARSOF units can provide CUL to the CA elements. The GSB/RSC can also act as a "plug-in" to the theater-level sustainment units on the ground. The deployed CA unit should establish a relationship early on with the 528th SB(SO)(A)'s ALE/ASPO cell, especially when there

MAINTENANCE SECTION

7-8. The mission of the 95th CA Brigade ground maintenance section is to provide quality maintenance for four battalions and one HQ on all CA ground support equipment and weapons in both CONUS and OCONUS.

RIGGERS SECTION

7-9. The 95th CA Brigade riggers provide personnel and cargo parachute support for the brigade. They provide a backfill for the SOTF as required for CA/USASOC airdrop and sling-load missions. In addition, they provide container delivery system airdrop and sling-load support as required to meet the 95th CA CMO and humanitarian assistance mission requirements in four theaters.

PLANNING AND PREPARATION CONSIDERATIONS

7-10. CA planners must address the planning considerations identified in Chapter 2. Two methodologies of planning are contingency planning and crisis action planning.

7-11. During contingency planning, CA units fully identify support requirements for OPLANs and CONPLANs in a SOR, down to the user level. In this way, the ASCC coordinates how to fulfill requirements from the support structure in the ASCC and prepares a support plan identifying support relationships. During contingency planning, the 528th SB(SO)(A) theater of operations ALEs assist the ASCC in conducting assessments or site surveys (Appendix D). When feasible, planners integrate these assessments into the theater of operations campaign plan to provide operational and logistics information for logistics preparation of the theater.

7-12. In crisis action planning, the requirements anticipated at the GCC's level dictate the amount of responsiveness and improvisation required in reactive, no-notice support and sustainment. Upon notification of mission requirements, CA units submit a revised SOR modifying initial logistics requirements. The use of assessment teams may not be practical during crisis action planning. In such cases, the relationship and coordination between the GCC, TSOC, 528th SB(SO)(A), and ALE are critically important.

STATEMENT OF REQUIREMENTS

7-13. The CA SOR is a critical source of information for the TSC and the ALE in their coordination and facilitation functions. The intent of the SOR process is to identify logistics needs early in the planning cycle. (Appendix E outlines the SOR format.) The TSOC ALE uses the GCC OPLAN in preparing his CONPLAN for inclusion in the mission order. This approach allows the ALE time to review required support before the CA mission unit submits the mission-tailored SOR. This review is especially critical in crisis action planning and short-notice mission changes.

ARMY HEALTH SYSTEM SUPPORT

7-14. Army CA units have medical personnel assigned to advise, evaluate, and coordinate medical infrastructure, support, and systems issues in foreign countries. Particular emphasis is on preventive medicine, sanitation, disease prevention, veterinary medicine, and prevention of zootoxin diseases. For sustainability of such services, CA units are dependent on the theater of operations AHS assets for most requirements. The CA battalion has assigned medical Soldiers and can provide limited AHS support to members of the unit in some mission profiles.

Chapter 7

STABILITY OPERATIONS

7-15. Each stability operation mission is unique and requires mission-specific analysis that develops a tailored sustainment force. Joint, international, and interagency activities add complexity to the sustainment system. CA forces may find themselves conducting operations outside a theater of operations support system because of geographic location. Preparation and submission of a SOR during these types of operations not only enhances the unit's requirements determination process, but also provides an opportunity to validate the theater of operations OPLAN requirements.

MAJOR OPERATIONS

7-16. A robust sustainment system that develops into a mature logistics infrastructure characterizes a protracted major operation. When the theater of operations support system is in place, it meets most ARSOF requirements. Logistics planners must focus on—

- *Initial entry.* They must determine the type of sustainment required, the number of days of accompanying supplies based on the time-phased force and deployment data, and the CA basing needs.
- *Buildup and integration.* They must coordinate and integrate CA logistics with the theater of operations support system before time-phased force and deployment data closure and as the system matures. In some cases, the theater of operations logistics infrastructure never achieves full maturity.
- *Redeployment.* As units start the redeployment phase, the ASCC ensures the tailoring (foreign nation support or contract) of the remaining support units to meet stay-behind CA support requirements.

Chapter 8

Military Information Support Operations Command

MISO units are characterized by their extraordinary ability to analyze and deal with complex politico-military problems. Their varied skills include techniques in persuasive, cross-cultural, and mass media communications; practical knowledge of social and behavioral psychology; cultural and situational awareness; and foreign language proficiency. These skills, unique to MISO, are proven keys to successful mission accomplishment and task attainment.

MISSION

8-1. MISO forces are organized and trained to conduct in-depth analyses of foreign target audiences, concentrating on their cultural, historical, political, social, economic, and religious characteristics, for the purpose of exploiting their psychological vulnerabilities. MISO's primary purpose is to modify the behaviors of the foreign target audiences to align with U.S. policy objectives. MISO forces are flexible and adaptable and are proficient in operating in widely diverse environments, conditions, and situations. MISO forces can operate in small autonomous teams or with other SO, conventional, or multinational units, or with other government agencies. MISO units provide a unique capability to other military forces. They are designed to meet the needs of conventional and other SOF commanders in all operational environments.

ORGANIZATION

8-2. The MISOC(A) is a provisional component subordinate command of USASOC (Figure 8-1, page 8-2). The unit is composed of a command group, HQ and staff, a separate ARSOF Media Operations Battalion, a Strategic Studies Detachment, and two ARSOF MISGs, each with three subordinate battalions. All battalions deploy in support of the GCCs, TSOCs, select joint force commanders, U.S. Ambassadors, other government agencies, and host and partner nation information efforts. To provide sufficient staff experience and expertise to plan, conduct, and support influence efforts, the MISOC(A) provides each TSOC and select USSOCOM component commands with a MIS planning and advisory team that is assigned to the MISOC(A) with duty at each TSOC and select USSOCOM component commands on a permanent basis. The MISOC(A) organizes, trains, equips, and deploys forces to the respective GCCs and TSOCs IAW applicable deployment and execution orders.

8-3. The MISOC(A) provides ARSOF MIS, to include advising, planning, and communicating, that ranges from area and target analysis, product development, and media production at the strategic and operational levels, to information collection, product distribution, and dissemination at the tactical level. ARSOF MISO provide a mission-tailored capability that combines skilled planners with technical assets and the flexibility to support TSOC security cooperation efforts in one part of the theater of operations while simultaneously providing direct support to other SOF core activities elsewhere. The organic media capabilities include fixed and deployable printing presses; television; multifrequency radio broadcasting stations; fixed and deployable audio, visual, and audiovisual production capabilities; and tactical loudspeaker dissemination. FM 3-53, *Military Information Support Operations*, includes additional information on MISO organizations.

Chapter 8

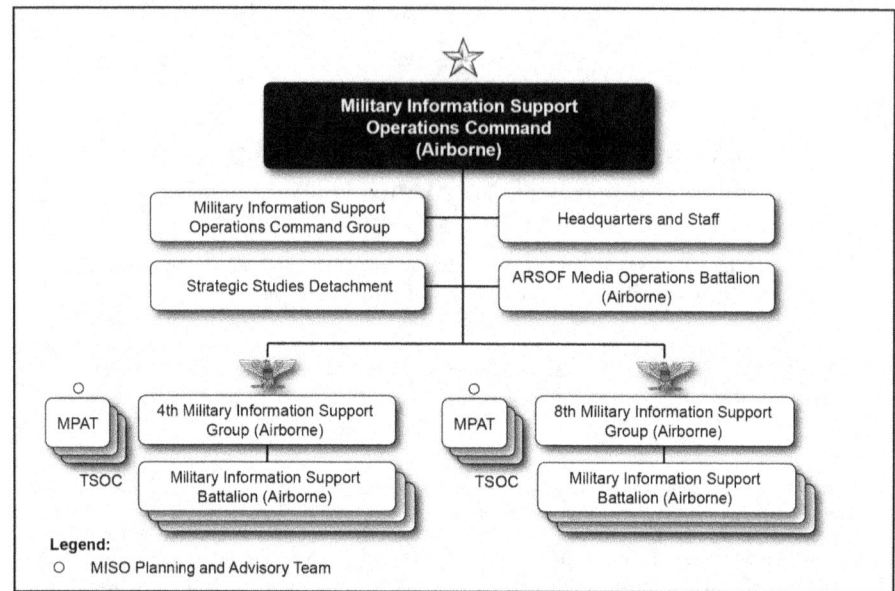

Figure 8-1. Military Information Support Operations Command (Airborne) organization

SUSTAINMENT

8-4. The challenge for sustainment personnel is to synchronize support activities with operational employment concepts. The logistics planner must plan a support concept for a joint MIS task force that operates from multiple bases, ranging from the CONUS through the communications zone to the JOA. The task becomes more complex by the requirement to derive common support from ARSOF, joint, and multinational sources. The following paragraphs outline the primary tasks for sustainment support to ARSOF MISO, the concept of support, responsibilities for support, and planning considerations.

PRIMARY TASKS

8-5. The logistician must be familiar with ARSOF and joint logistics. He must also be knowledgeable in securing support from multinational or HN sources.

8-6. The SOR is the key to securing responsive support. Based upon current operational plans, the SOR is updated with receipt of a warning order from the applicable theater of operations through USSOCOM and during contingency planning. Like the joint targeting coordination board in the operational planning process, the SOR is recognized by Army and sister Service logisticians. The SOR is a powerful tool that logisticians must master and use.

8-7. ARSOF MISO units are based in the CONUS and operate IAW the Army's strategy of using force projection. USASOC has organized and stationed its ARSOF sustainment organizations and activities IAW the Army's concept of force projection. This change allows MISO units to integrate organic logistics and health protection elements within the ASCC support structure for sustainment to deployed ARSOF MISO units to meet this challenge.

CONCEPT OF MILITARY INFORMATION SUPPORT SUSTAINMENT

8-8. As a result of the reorganization of USAR MISO units to the conventional Army, active MISO units, primarily, will support SOF. ARSOF doctrine is addressing the implications of the restructuring of MISO forces, which includes logistical support. In all likelihood, the TSOCs may be providing support to ARSOF MISO units.

8-9. The MISG or regional MIS battalion S-4 has the staff lead for logistics planning and execution. When not task-organized for an operational mission, the group S-4 is the senior logistics officer, and the USASOC Deputy Chief of Staff for Logistics is the higher logistics authority. When task-organized for an operational mission, the group S-4 coordinates with the HQ having control to establish the logistics support relationships. The group S-4 should also be coordinating with the deployed 528th SB (SO)(A) ALE and ASPO cell for any coordination that may need to take place between the MISO units, the ASCC theater of operations support elements, and the TSOC. The S-4 must arrange for continuity of logistics support during the transition between USASOC and the change of operational control to the GCC.

8-10. The concept of MISO support must continuously focus on maintaining and sustaining the operational strength of the deployed units. MISO require extensive resources in analysis, multimedia production and dissemination, and assessment. The operational demands often exceed the capabilities of the MISO unit and the supported unit, necessitating the use of contracting to meet mission requirements. Contracting is done routinely to acquire services for sustainment of units and equipment; development, delivery, and assessment of MISO products and actions; and to fulfill other mission requirements.

SUSTAINMENT RESPONSIBILITIES

8-11. Unless otherwise directed by the Secretary of Defense, the military departments and Services continue to provide the logistics and administrative support of Service forces assigned or attached to joint commands, including ARSOF MISO units. Elements with support responsibilities include the ASCC, USASOC, and TSOC.

8-12. As prescribed by 10 USC, USASOC retains responsibility for logistical support of ARSOF. The ASCC provides the necessary support for the Army forces assigned to a combatant command. With assistance from the 528th SB(SO)(A) and the ALE, the ASCC develops the theater of operations support plan that includes sustainment of MISO units by theater of operations logistics organizations. In ISBs, the ASCC, with the assistance of the 528th SB(SO)(A) and ALE, also provide support to ARSOF and other Army forces, as directed. MISO units have some key differences that affect the type of support required for RSOI and sustainment. The following conditions occur often enough that they must receive special consideration during logistics planning:
- Supply distribution is the key consideration for deployed ARSOF MISO units located in isolated and austere locations.
- ARSOF MISO units have significant amounts of unique equipment that require support through SO-peculiar logistics channels.
- ARSOF MISO units have extensive and unique contractual requirements.
- ARSOF MISO units have extensive and unique requirements for financial support.

8-13. The 528th SB(SO)(A), ALE, and the ARSOF MISO command structure monitor ongoing logistics support to MISO forces and provide the initial support that may not be available from the ASCC. The following organizations perform these functions:
- The 528th SB(SO)(A) assists in the planning and coordination of ARSOF MISO logistics requirements. The brigade may coordinate Army-common sustainment and SOF-peculiar support by working with both the TSOC and ASCC to ensure the MISO-developed plans are integrated into the ASCC concept of support for the theater of operations. The brigade may also attach logistics liaison officers when its sustainment operations are expected to require complex multi-Service, interagency, and contractual support.

- The ALE has a coordination cell with the ASCC staff. It provides SO staff expertise and coordinates access to the support infrastructure. It ensures ARSOF MISO requirements are included in the support plan. It also provides the capability for deploying ARSOF MISO units to gain access to the theater of operations Army support structures upon arrival in the theater of operations.
- The 528th SB(SO)(A), when required, provides limited direct support to ARSOF MISO units. It provides support from the early arrival and employment of units until the theater of operations support structure capability can take over. The brigade is capable of deploying anywhere in the world to provide early support. It provides support only until the theater of operations support structure is established and capable of meeting ARSOF MISO requirements. Once the theater of operations support structure is in place, the brigade elements prepare to redeploy in preparation for other contingencies.

8-14. The TSOC supports ARSOF MISO units for any MISO-peculiar requirements the 528th SB(SO)(A) or ASCC identifies as a shortfall. The TSOC also works closely with the combatant command staff, the theater of operations ASCC, and MISO logisticians to convey the MISO requirements, depending upon the theater of operations organizational structure.

8-15. The TSOC and ARSOF MISO logisticians coordinate with the ASCC to develop plans and subsequent orders to implement directives the ASCC issues to support the ARSOF MISO units assigned to the combatant command. The TSOC, in conjunction with the ARSOF MIS battalion S-4, advises the ASCC commander on the appropriate logistics command and support relationships for each ARSOF MISO mission. The ALE keeps the 528th SB(SO)(A) and USASOC informed of the status of ASCC supporting plans.

PLANNING CONSIDERATIONS

8-16. When circumstances allow, ARSOF MISO units and the 528th SB(SO)(A) apply contingency planning to identify support requirements in OPLANs and CONPLANs. Some of these requirements are based upon an established set of planning assumptions or SORs, down to the user level. In this way, the 528th SB(SO)(A) and the ASCC coordinate how to fulfill requirements from the support structure in the theater of operations Army. In crisis action planning, the requirements anticipated at the GCC's level dictate the amount of responsiveness and improvisation required to provide reactive, no-notice support and sustainment. Actual circumstances may dictate the modification of preplanned requirements, or they may generate new requirements unanticipated during the contingency planning process.

STATEMENT OF REQUIREMENTS

8-17. The SOR is a critical source of information the 528th SB(SO)(A), ASCC, and the ALE need in their coordination and facilitation functions. (Appendix E outlines the SOR format.) The intent of the SOR process is to identify logistics needs early in the planning cycle. The TSOC J-4 uses the ASCC operational plan in preparing his CONPLAN for inclusion in the mission order. This approach allows the TSOC ALEs time to review required support before the ARSOF MISO unit submits the mission-tailored SOR. This review is especially critical in crisis action planning and short-notice mission changes.

SPECIAL MAINTENANCE CONSIDERATIONS

8-18. ARSOF MISO units normally deploy with a limited organizational maintenance capability. They may need to obtain field and sustainment maintenance from the theater of operations ASCC elements (TSC, ESC, CSSB, and brigade support battalion) for Army-common equipment. MISO-unique nonstandard equipment may need to be coordinated through the supporting theater of operations contractual activity, the TSOC with coordination by the 528th SB(SO)(A), or reachback to home station (CONUS). The reachback capability is a unique and focused means to evacuate, repair, and replace SOF-unique equipment.

8-19. Tactical ARSOF MISO units are attached to the forces they support. Normally, in a mature theater of operations, they receive maintenance support from the supported unit. Supported units are normally unable

Military Information Support Operations Command

to provide maintenance support for loudspeakers and other MISO-unique equipment. Tactical MISO forces are usually dependent upon their own electronic maintenance shop for support. Contractors are usually attached for maintenance purposes.

ARMY HEALTH SYSTEM SUPPORT

8-20. ARSOF MISO units do not have any organic AHS assets. They are entirely dependent upon the supported unit for all aspects of AHS support.

This page intentionally left blank.

Chapter 9
Army Health System Support

This chapter addresses the AHS, which consists of HSS, a component of the sustainment warfighting function, and FHP, a component of the protection warfighting function. These two components, though separated into different warfighting functions, must be considered equally during SO planning and execution.

The provision of AHS support for ARSOF is challenging, with these forces having few organic support assets and routinely entering austere theater of operations before adequate support structures are established. SO have limited organic AHS support resources and require support from conventional forces medical assets for the majority of their health care requirements. The planner must comprehensively anticipate, coordinate, and synchronize resources to ensure prompt medical intervention is available without compromising the SO mission. In planning and coordination for medical support (internal and external), the planner must analyze operational mission planning factors against the ten medical functional areas to determine materiel and equipment requirements and develop the mission-specific medical support plan. More in-depth AHS support information is in FM 4-02.12, *Army Health System Command and Control Organizations*, FM 4-02.43, and ATTP 4-02.

ARMY SPECIAL OPERATIONS FORCES MEDICAL CONSIDERATIONS

9-1. To provide effective AHS support, the planners must understand how the OPLAN is to be executed, what unique challenges will result, and the emphasis on asymmetrical and unconventional warfare. ARSOF conduct the full range of military operations where AHS planners must focus on and understand the mission. Appendix F discusses the health threat and medical intelligence.

SPECIAL OPERATIONS FORCE GENERATION RESET CYCLE (POSTDEPLOYMENT)

9-2. During the SOFORGEN Reset (Postdeployment) Cycle, the senior medical officer is responsible for programming (synchronizing, coordinating, scheduling, and acquiring) unit medical tasks across five HSS domains: medical readiness, medical materiel, medical training, medical enablers, and behavioral health.

Medical Readiness

9-3. The focus during the Reset phase is postdeployment readiness and includes tasks associated with immunizations and health screenings, postdeployment health assessments/reassessments, identifying and tracking deployment-limiting profiles, dental readiness, and women's health considerations (if applicable).

Medical Materiel

9-4. The focus during the Reset phase is assessing Class VIII requirements and planning to rectify any shortcomings. Tasks include scheduling maintenance and calibration of medical equipment; inventorying sets, kits, and outfits; ordering shortages; and identifying and replacing potency and dated materiel.

Chapter 9

Medical Training

9-5. The focus during the Reset phase is programming all medical training that must occur before the next deployment. Tasks include forecasting veterinary support for live tissue training/combat trauma modules and managing the credentialing program, career schools, and skills sustainment courses for medical personnel.

Medical Enablers

9-6. The focus during the Reset phase is identifying and programming service-provided-capability medical units that are needed to mitigate risk based on course-of-action analysis and the military decisionmaking process. Upon receipt of an order and entering the initial planning phase, planners must identify any medical enablers required and should attempt to establish contact with that unit, providing initial training guidance.

Behavioral Health

9-7. The focus during the Reset phase is postdeployment behavioral health. Tasks include face-to-face behavioral health screenings IAW current HQDA guidance and any other medical requirements associated with post-traumatic stress disorder and traumatic brain injuries.

SPECIAL OPERATIONS FORCE GENERATION TRAIN-READY CYCLE (PREPARING FOR DEPLOYMENT)

9-8. During the SOFORGEN Train-Ready (Preparing for Deployment) Cycle, the planners must conduct operational medical planning and geographical/environmental medical threat analysis prior to all deployments to determine materiel quantities and specific additional support requirements for that mission. Medics must fully understand the SOF core tasks and the operational, tactical, and geographical constraints associated with those missions.

Medical Readiness

9-9. Focus during the Train-Ready phase is predeployment readiness and includes tasks associated with achieving 100-percent compliance with applicable personnel policy guidance tasks (for example, immunizations, hearing exams, resolving deployment-limiting profiles, and so on). Complying with the personnel policy guidance may require interface with the combatant command and TSOC surgeons to work individual exceptions to policy/medical waivers.

Medical Materiel

9-10. The focus during the Train-Ready phase is bringing all medical sets, kits, and outfits to fully-mission-capable status. Units will receive/pack/utilize Class VIII for training, fill shortages, receive new equipment fieldings, and train on any new equipment received. Units will draw narcotics and functionally pack their medical equipment as deployment approaches.

Medical Training

9-11. The focus during the Train-Ready phase is completing the training that was programmed in the Reset phase. Tasks include actual live tissue training/combat trauma modules, ensuring medical personnel remain credentialed, and directly supporting training events and non-trauma modules, either formally or informally, per the unit surgeon's discretion. Special focus is required for ARSOF medics (SF medical sergeants, SOCMs, CA medical sergeants), who must complete SOCM skills sustainment training and a Medical Proficiency Training (MPT) rotation, to maintain their skill sets. All assigned or attached Servicemembers must complete Tactical Combat Casualty Care (TCCC) prior to deployment.

Medical Enablers

9-12. The focus during the Train-Ready phase is formalizing habitual relationships with medical enablers. Those units should participate in all premission training and culminating training exercises.

Behavioral Health

9-13. The focus during the Train-Ready phase is predeployment behavioral health readiness. Tasks include online training and certifications for traumatic brain injuries and post-traumatic stress disorder. The minimum behavioral health-related requirement is completing a baseline Immediate Post-Concussion Assessment and Cognitive Testing (ImPACT) for assigned or attached Servicemembers prior to deployment.

SPECIAL OPERATIONS FORCE GENERATION AVAILABLE CYCLE (DEPLOYED)

9-14. During the SOFORGEN Available (Deployed) Cycle, the now deployed planner must continually refine the medical support plan by conducting medical support analysis and modify the plan as theater of operations medical support assets become more or less available. Although augmentation of ARSOF medical resources may be required for a number of types of missions, the most likely mission where medical company resources would be employed in direct support or general support is the FID mission. The following SOF characteristics must be factored into the plan.

Small Units and Austere Force Health Protection Capability

9-15. ARSOF will require support on an area basis. Unit locations (geographical factors, time-distance factors) may require collocation of assets.

Operations in a Joint, Allied, and Coalition Environment

9-16. Operations in these environments require officers and NCOs to have a thorough knowledge of their sister Services/allied/coalition forces' HSS capabilities, limitations, organization, procedures, and national caveats.

Remote Operating Areas and Long Evacuation Routes

9-17. ARSOF units operate at a long distance from airfields suitable for evacuation. ARSOF often operate in areas that impede evacuation by rotary-wing aircraft or where aviation assets are not available. This places a premium on the early application of trauma management and casualty stabilization.

Medical Evacuation, Medical Regulating, and Casualty Tracking

9-18. Medical regulating and casualty tracking requires an understanding of ARSOF missions and the limited availability of replacements. Leaders must account for sensitive equipment and/or documents if the casualty still possesses them when evacuated.

Working With Indigenous Populations

9-19. Additional considerations include local medical infrastructure, theater of operations medical rules of engagement (combatant/noncombatant) regarding rendering of health care and health education initiatives, cultural beliefs, and effects on the population. Specific attention must be given to FHP measures to protect the Soldier against endemic/epidemic disease threats.

HEALTH SERVICE SUPPORT

9-20. HSS encompasses all services performed, provided, and arranged by the Army Medical Department to promote, improve, conserve, or restore the mental and physical well-being of Army personnel as directed in other Services, agencies, and organizations.

Chapter 9

9-21. The unit surgeon, at all levels of ARSOF, is responsible for planning, coordinating, and synchronizing HSS functions and missions, including the coordination necessary to ensure that medical support is available when requirements exceed the organic capabilities of his unit. The unit surgeon is responsible for determining medical requirements and providing oversight for the requisition, procurement, storage, maintenance, distribution management, and documentation of medical supplies and equipment, as well as a host of other HSS tasks.

9-22. Research specialists are dedicated to researching and compiling medical threat information in all foreign countries and disseminating this information to all deploying SO elements. The USSOCOM medical intelligence section is the interface between SO and the National Center for Medical Intelligence. Medical intelligence maintains comprehensive classified and unclassified hard-copy and electronic databases in support of SO deployments for training and security assistance commitments. It also maintains extensive liaison with intelligence and medical networks within USSOCOM, National Center for Medical Intelligence, Defense Intelligence Agency, and other agencies. This section compiles new critical elements of information received from teams operating in the field for dissemination in future deployments. After-action reports containing medical information have proven to be critical in planning operations. The preventative medicine branch and medical intelligence section work together in recognizing the threat and recommending countermeasures to this threat.

MEDICAL TREATMENT

9-23. Role 1 treatment is organic to all ARSOF units and is provided by SOCMs, SF medical sergeants, CA medical sergeants, or physicians and physician assistants.

9-24. Limited Role 2 capability exists within the SFGs and the 528th SB(SO)(A). Figure 9-1, page 9-5, has additional information on the Role 2 capability.

9-25. There is no surgical capability organic to any ARSOF unit. ARSOF rely on area support or augmentation (for example, forward surgical teams or area support medical companies) from the conventional forces or sister Services for surgical capability. It is imperative that these external surgical units have an acceptable level of familiarity with SOF operations, as their employment in support of SO places unique demands on their capability.

MEDICAL EVACUATION

9-26. Planners must understand the difference between CASEVAC and MEDEVAC, as these terms mean different things to medical personnel. The primary difference between a CASEVAC and a MEDEVAC is the method of evacuation. MEDEVAC uses a standardized and dedicated vehicle providing en route care, whereas CASEVAC uses nonstandard and nondedicated vehicles that do not provide en route care. CASEVAC is used to transport casualties that are in dire need of evacuation from the battlefield and do not have time to wait on a MEDEVAC, or where a MEDEVAC is unable to get to the casualty. (Appendix G discusses MEDEVACs.)

9-27. Although the SO health care provider receives enhanced medical training exceeding the level and scope found in conventional forces, he depends heavily on the conventional AHS to conserve the combat strength of ARSOF (particularly in the area of MEDEVAC where ARSOF do not have a dedicated system). Ideally, MEDEVAC for ARSOF personnel should follow the doctrinal flow sequence. The ARSOF planner must be innovative and follow the tenets of immediate far forward stabilization. He directs evacuation to the appropriate military treatment facility when the condition of the patient warrants it, with whatever means of transportation are available.

Army Health System Support

Medical Planner is responsible for coordinating all of these functions.
Instructions -- Find the capability on the left and follow it to the right to determine who can provide the asset/resource.
"√" Indicates organic capability.
"Ltd" Indicates limited capability with organic assets.
"Role" is specific to Army doctrine only for the purposes of this table.
THOR3 - Tactical Human Optimization Rapid Rehabilitation Reconditioning

Category	Capability	Role 1 ARSOF						Role 2			Role 3		Role 4	
		SF and Ranger Team/Company	SF and Ranger Battalion	SF Group/CJSOTF	528th SB(SO)(A)	Self/Buddy Aid Combat Lifesaver	Battalion Aid Station	Brigade Support Medical Company (BSMC)	Area Support Medical Company (ASMC)	Forward Surgical Team (Can Attach to SOF)	Combat Support Hospital	Expeditionary Medical Support (Air Force)	Landstuhl Regional Medical Center (LRMC)	CONUS Medical Treatment Facility (MTF)
MEDEVAC Regulation	Locate, Collect, Transport/En Route Medical Care	√	√	√	√	√	√	√	√		√	√		
	Locate Available MTFs/Coordinate Transport	√	√	√	√	√	√	√	√		√	√		
	CASEVAC (Nonstandard Platforms [Sea/Land/Air])	√	√	√	√	√	√							
	Air Force Air (Coordinate Air Force Air Liaison Officer)										√	√	√	√
MEDEVAC Treatment	Tactical Combat Casualty Care (TCCC)	√	√	√	√	√	√	√	√					
	First Responder/First Aid/Immediate Lifesaving	√	√	√	√	√	√	√	√					
	Emergency Medical Treatment	√	√	√	√	√	√	√	√	√	√	√	√	√
	Sick Call	Ltd	√	√	√			√	√		√	√	√	√
	Advanced Trauma Management	√	√				√	√	√	√	√	√	√	√
	Routine Physical Therapy/Rehabilitation/THOR3						√	√		√	√	√	√	√
	Ultrasound	Ltd	Ltd	Ltd							√	√	√	√
	X-ray		Ltd	Ltd	Ltd			Ltd	Ltd		√	√	√	√
	Initial Wound Surgery	Ltd	Ltd	Ltd						√	√	√		
	Resuscitative Wound Surgery									√	√	√		
	Postoperative Treatment										√	√	√	√
Hospital	Patient Hold		Ltd	Ltd	Ltd			√	√		√	√	√	√
	Definitive Rehabilitative/Convalescent Care												√	√
Medical Logistics (Class VIII)	Requesting, Receiving, Issuing, Storing Class VIII	√	√	√	√								√	√
	Medical Equipment Maintenance/Repair			√	√			√	√				√	√
	Prescription Optical Lens Fabrication			Ltd				√	√				√	√
	Blood Storage/Distribution	Ltd	Ltd	Ltd	Ltd			Ltd	Ltd	Ltd	√	√	√	√
Preventive Medicine (PM)	Disease and Nonbattle Injury Prevention	√	√	√	√	√		√	√		√	√	√	√
	Weapons of Mass Destruction (WMD)	√	√	√	√			√	√					
	Communicable Diseases	√	√	√	√			√	√					
	Environmental/Occupational Health Threats	√	√	√	√			√	√					
	Waste (Human, Hazardous, Medical Disposal)	√	√	√	√			√	√					
	Immunizations/Medical Readiness	√	√	√	√		√	√	√		√	√	√	√
Vet Services	Food Safety and Inspection	√	√	√	√									
	Multipurpose Canine Support	Ltd	Ltd	√	√									
	Animal Medical Care/Husbandry (Canine/Livestock)	√	√	√	√									
Dental	Dental Care	Ltd	Ltd	√	√			√	√		√	√	√	√
	Resuscitative Maxillofacial Surgery										√	√	√	√
Behavioral Health (BH)	Personal Issues/Resiliency	Ltd	√	√	√						√	√	√	√
	Combat Operational Stress Prevention	Ltd	√	√	√						√	√	√	√
	Traumatic Brain Injury/Impact	Ltd	√	√	√						√	√	√	√
	Post-Traumatic Stress Disorder	Ltd	Ltd	√	√						√	√	√	√
Lab Services	Assess Disease Processes	Ltd	√	√	√			Ltd	Ltd		√	√	√	√
	Monitor Efficacy of Medical Treatment	√	√	√	√			Ltd	Ltd		√	√	√	√
	Confirmatory Identification of Biological Threat Agents	√	√	√	√			Ltd	Ltd		√	√	√	√
	Communications and Computers	√	√	√	√		√				√	√	√	√

Figure 9-1. ARSOF health service support capabilities

9-28. MEDEVAC of ARSOF casualties is an operational matter. That is, it must reflect the commander's concept of the operation, and can only succeed when the planner integrates the MEDEVAC plan with the tactical plan and logistics airflow. The SO planner must constantly coordinate with the battalion or group operations and logistics sections to obtain up-to-date information of opportune transportation assets to be used for evacuation.

MEDICAL REGULATING

9-29. Medical regulating is not an ARSOF function since units are neither staffed nor equipped to accomplish this mission. It is essential that surgeons at all levels understand how patients are regulated within and between theaters of operations, and how they can track them through the system. Planners must ensure that SO have their own evacuation policy to allow return of critical SO medical occupational specialties to their units instead of being evacuated out of the theater of operations. Intra-theater of operations medical regulating of ARSOF is normally an Army function. The transfer of patients from a Role 3 Army military treatment facility to a Role 4 facility is a theater of operations (or joint) function; therefore, intratheater of operations and intertheater of operations medical regulating is usually a function of the Theater Patient Movement Requirement Center and Global Patient Movement Requirements Center.

9-30. The medical regulating plan must be integrated with the ARSOF operational and logistic plan. During sustained SO missions, the TSOC cannot afford to lose the services of low-density SO skilled Soldiers who become casualties. Every effort must be made to preserve SO capabilities and combat power in the theater of operations. A determination must be made as to who can be treated and returned to duty at hospitals within the GCC or theater of operations. As an exception to the theater of operations evacuation policy, the combatant command may retain injured or wounded SO Soldiers in theater of operations where they can be returned to limited duty without jeopardizing their recovery and health. There they can assume the support duties performed by other SO Soldiers, freeing the latter for operational duties. Due to the shortage of SO personnel and criticality of MOS skills within the theater of operations, an exception to the theater of operations evacuation policy may be required for SOF to facilitate the return to duty of these patients within the operational environment.

HOSPITALIZATION

9-31. SOF do not have an organic Role 3 capability and rely on the theater of operations hospitalization system for their patients. In an underdeveloped operating environment, SOF may rely on the HN to provide hospitalization. When the sensitivity of a particular mission dictates strict operational security, the SO commander must coordinate with the TSOC surgeon to establish facilities capable of handling patients on a classified basis.

PREVENTIVE MEDICINE SERVICES

9-32. A major shortfall of SOF HSS is the lack of organic preventive medicine personnel for extensive area support (such as aerial spraying and larviciding). Although SF medical sergeants are trained in the basics of preventive medicine, SO units have limited assets and capabilities to plan, coordinate, and supervise preventive medicine programs to the extent that is required. Given the nature of SOF operations, which places personnel at serious risk for disease and environmental injury, a full-time preventive medicine commitment may be required, requiring the use of area support medical units. The preventive medicine NCOs in the SF battalions and the 528th SB(SO)(A) provide technical assistance to the unit field sanitation teams and advise the commander on the control measures required to protect the force. Additionally, environmental science officers provide subject-matter expertise and are assigned in most USASOC units.

DENTAL SERVICES

9-33. A dental officer (area of concentration [AOC] 63A) and dental technician (MOS 68E) are assigned to the SFGs, the 95th CA Brigade, and to the 528th SB(SO)(A) and are capable of providing emergency or sustaining dental care. The SF medical sergeants have limited dental training and equipment and can provide only emergency dental care.

VETERINARY SERVICES

9-34. ARSOF have veterinary service personnel assigned and/or attached at various command levels. When veterinary services are required in more than one location, or when the requirements exceed two deployed SOTFs, veterinary support requires external augmentation. Veterinary personnel can perform the majority of the food source inspection mission and service the units' multipurpose canines. Additionally, they—

- Work with indigenous military assets and allied or foreign governmental agencies.
- Assist in planning and executing population and resource control, civic action, and other security, development, and stability programs.
- During military and paramilitary operations, assist in planning and executing civic action, humanitarian assistance, and other programs designed to expand the government's legitimacy within contested areas.
- Provide estimates and data on the resources essential to build an effective infrastructure for civil health and agricultural administration and operations.
- Assess available infrastructure to support combat forces.
- Support and coordinate humanitarian and disaster relief in coordination with other DOD elements, other U.S. Government agencies, foreign and HN authorities, and international relief organizations.
- Assist in the planning and coordination of noncombatant evacuation operations in the areas of food supply and privately owned pet (dogs and cats) evacuation.
- Assist in coordinating the use of local HN resources such as maintenance of veterinary care facilities. They provide and conduct public health, veterinary preventative medicine, and civil defense operations in conjunction with the local agencies.

COMBAT AND OPERATIONAL STRESS CONTROL

9-35. Combat fatigue cases should be managed as far forward as possible to preclude unnecessary loss of personnel, hasten returns to duty, and prevent overburdening the MEDEVAC system. ARSOF do not have organic combat and operational stress control teams; however, most SO units have organic Behavioral Health Officers (AOC 67D) and Behavioral Health Technicians (MOS 68X).

HEALTH SERVICE LOGISTICS

9-36. The medical sections of all SOF units maintain a Class VIII (medical supply) basic load to support initial operations. The SFG, 528th SB(SO)(A), and battalion medical sections are the only SO elements with organic medical supply personnel. No SO unit has an organic medical equipment maintenance capability. Units receive routine medical logistics support through their supporting medical battalion (multifunctional). This support includes Class VIII supplies, oxygen, resuscitative fluids production, optical fabrication, medical equipment maintenance support, and blood management. To fill operational requirements in support of UW or FID, medical supply personnel at the SOTF level requisition bulk Class VIII supplies directly from the supporting medical battalion (multifunctional) or installation medical logistics activity. ARSOF may also supplement their medical logistics effort with foreign national medical supplies, particularly during UW operations, if approved by the Food and Drug Administration and theater of operations surgeon.

MEDICAL LABORATORY SERVICES

9-37. The SFODA and the 528th SB(SO)(A) are the only units with laboratory capability, but this capability is extremely limited. The SF medical sergeant (MOS 18D) is trained to provide basic clinical laboratory tests and procedures in support of UW or FID missions. Role 3 laboratory support is obtained on an area support basis.

FORCE HEALTH PROTECTION

9-38. FHP consists of the measures to promote, improve, or conserve the mental and physical well-being of Soldiers, to include those that enable a healthy and fit force, prevent injury and illness, and protect the force from health hazards.

HEALTHY AND FIT FORCE

9-39. USASOC's Tactical Human Optimization Rapid Rehabilitation Reconditioning (THOR3) program is part of a larger USSOCOM mission-specific human performance program (HPP). As tactical athletes, ARSOF warriors must maintain peak performance in addition to general health and fitness. At the elite tactical athlete level, there are major distinctions in programs designed solely for fitness versus human performance optimization. THOR3 increases combat performance and effectiveness, prevents injuries, improves health and longevity, and facilitates rapid return to duty.

CASUALTY PREVENTION

9-40. Education and thorough indoctrination to the risks and surveillance procedures are continually required to safeguard the health and readiness status of the operational force.

CASUALTY CARE AND MANAGEMENT

9-41. Casualty care and management within ARSOF consists of the measures to provide first response, forward resuscitative surgery, hospitalization, en route care, and definitive care. The combination of these measures ensures ARSOF Soldiers receive the best and most efficient care and management in the theater of operations.

First Response

9-42. A combination of organic, direct support, and general support resources are required to effectively accomplish the HSS mission. ARSOF are characterized by an austere structure and a limited number of medical personnel with enhanced medical skills, including emergency medical technician, advanced trauma management, advanced trauma life support, and limited veterinary and dental care. In addition, ARSOF support units have flight surgeons and medics who are qualified to provide Echelon I care.

Forward Resuscitative Surgery

9-43. AHS planning for missions where the risk of penetrating trauma is high and further complicated by time and extended distance requires extensive planning and coordination. The capability for medical personnel and units that can perform resuscitative surgery and invasive stabilization must be provided regardless of the ARSOF unit involved.

Theater Hospitalization

9-44. As addressed in paragraph 9-31, SOF do not have an organic Role 3 capability and rely on the theater of operations hospitalization system for their patients.

En Route Care

9-45. Planners must ensure there is adequate MEDEVAC capability, both intertheater of operations and intratheater of operations. If SOF assets are used, as will probably be the case in intra-theater of operations evacuation, medical assets must be on board to provide medical care en route. Early coordination must be made with theater of operations USAF assets or supporting SOA assets to ensure timely intertheater of operations evacuation capability.

Army Health System Support

Definitive Care

9-46. The planner must be involved as early as possible in the planning process; the FHP planner must produce a straightforward plan without complication to ensure a continuum of care to the full range of ARSOF operational environments that covers point of injury (POI) through Role 4 CONUS-based hospitals.

9-47. The ARSOF medical planner must plan for all HSS functional support areas regardless of the mission or the operating environment. HSS planning for ARSOF involves numerous considerations that do not apply to conventional forces. ARSOF units and personnel operate over a wide area and in isolated and austere locations with limited HSS. The medical plan (normally an annex to the OPLAN/OPORD) must be in sufficient detail to designate specific area support responsibilities provided to ARSOF units. A well-developed HSS annex of the OPLAN/OPORD ensures all medical functional support is available to support ARSOF personnel when required or requested. Direct support from conventional medical units may not be feasible during certain operations. Support may be required for ARSOF units/personnel following the completion of the mission. ARSOF units must coordinate support from theater of operations medical units. To enable this, the planner must develop a comprehensive plan and thoroughly coordinate and update the plan with the supporting medical HQ. In ARSOF operations where medical units are not available in the operating environment, tasking for specific HSS requirements and capabilities are critical. For planning purposes, Figure 9-2 provides organic and nonorganic AHS support by organization.

Functional Area	SFODA	SFODB	SF Bn	SFG	528th SB(SO) (A)	BCT	MMB	FST	CSH	Theater of Operations
Evacuation:										
CASEVAC Platforms (Air/Land/Sea)	X	X	X	X	X	X	X	X	X	X
MEDEVAC (Air)										X
MEDEVAC (Ground)						X	X			X
Treatment:										
Sick Call/Trauma	X	X	X	X	X	X	X	X	X	X
Surgery								X	X	X
Hospitalization									X	X
Patient Holding	LTD	LTD	LTD	LTD	LTD	LTD	LTD	LTD	X	X
Medical Logistics				LTD	LTD	X	X	X	X	X
Preventive Medicine			X	X	LTD	X	X	X	X	X
Veterinary Services				LTD	X					X
Dental Services				LTD	LTD	X	X			X
Behavioral Health				X					X	X
Laboratory Services					LTD	X	X	LTD	X	X
Medical Communications and Computers	LTD	LTD	LTD	X	LTD	X	X		X	X

NOTES: "LTD" indicates minimal organic capability; "X" indicates the element is capable of providing the function.

BCT   Brigade Combat Team   FST   Forward Surgical Team
Bn   Battalion   MMB   Multifunctional Medical Battalion
CSH   Combat Support Hospital   SFODB   Special Forces Operational Detachment B

Figure 9-2. Organic and nonorganic Army Health System support

This page intentionally left blank.

Chapter 10
Personnel Services Support

Personnel services support within ARSOF is limited to those services provided by personnel assigned to each unit. ARSOF units are reliant upon the GCC within the theater of operations for personnel services support. Personnel services support consists of five major subfunctions: human resources support, financial management, legal support, religious support, and band support. This chapter provides only an overview of those personnel services that may be available; more detailed information on personnel services support is in ADRP 4-0.

HUMAN RESOURCES SUPPORT

10-1. Human resources support is the aggregate of systems and services designed to provide and support Soldiers. Human resources support is important to maximizing operational reach and endurance. Each includes major functional elements and all are covered below (FM 1-0 includes more information). Human resources support encompasses four major categories:
- Manning the force.
- Human resources services.
- Personnel support.
- Human resources planning and staff operations.

MANNING THE FORCE

10-2. Manning the force involves personnel readiness of the force, maintaining accountability of the force, and management of personnel information. Manning combines anticipation, movement, and skillful positioning of personnel assets. It relies on the secure, robust, and survivable communications and digital information systems of emerging technologies that provide the common operational picture, asset visibility, predictive modeling, and exception reporting.

Personnel Readiness Management

10-3. The purpose of personnel readiness management is to distribute Soldiers to units based on documented requirements, authorizations, and predictive analysis to maximize mission preparedness and provide the manpower needed to support unified land operations.

Personnel Accountability

10-4. Personnel accountability plays a critical role in deployed operations and relies on timely, accurate, and complete duty status and location of personnel at all times. Personnel accountability is the process for recording by-name data on Soldiers when they arrive and depart from units; when their location or duty status changes (such as from duty to hospital); or when their grade changes. Activities also include the reception of personnel, the assignment and tracking of replacements, return-to-duty and rest and recuperation personnel, and redeployment operations.

Chapter 10

Strength Reporting

10-5. Strength reporting is a numerical end product of the accounting process, achieved by comparing the by-name data obtained during the personnel accountability process (faces) against specified authorizations (spaces or, in some cases, requirements) to determine a percentage of fill.

Personnel Information Management

10-6. Personnel information management encompasses the collecting, processing, storing, displaying, and disseminating of relevant information about Soldiers, units, and civilians. Personnel information management is the foundation for conducting or executing all human resources functions and tasks.

HUMAN RESOURCES SERVICES

10-7. Human resources services encompass casualty operations and essential personnel services to maintain Soldier readiness and sustain the human dimension of the force. The following is a discussion of casualty operations and essential personnel services.

Casualty Operations Management

10-8. The casualty operations management process includes the recording, reporting, verifying, and processing of information from unit level to HQDA. It also involves notifying appropriate individuals and assisting family members.

Essential Personnel Services

10-9. Essential personnel services provide Soldiers and units timely and accurate personnel services that efficiently update Soldier status, readiness, and quality of life, and allow Army leadership to effectively manage the force. Essential personnel services include actions supporting individual career advancement and development, proper identification documents for security and benefits entitlements, recognition of achievements, and service performance. It also includes personal actions such as promotions, reductions, evaluations, military pay, leave and pass, separations, and line-of-duty investigations.

PERSONNEL SUPPORT

10-10. Personnel support encompasses command interest/human resources programs; morale, welfare, and recreation; and retention functions. Personnel support also includes substance abuse and prevention programs, enhances unit cohesion, and sustains the morale of the force.

Postal Operations

10-11. Postal operations and services have a significant impact on Soldiers, civilians, and their families. Efficient postal operations are necessary and require significant logistics and planning for issues such as air and ground transportation, specialized equipment, secured facilities, palletization crews, mail handlers, and others. Postal services also include selling stamps; cashing and selling money orders; providing registered (including classified up to Secret), insured, and certified mail services; and processing postal claims and inquiries.

Morale, Welfare, and Recreation and Community Support

10-12. Morale, welfare, and recreation and community support provide Soldiers, Army civilians, and other authorized personnel with recreational and fitness activities, goods, and services. The morale, welfare, and recreation network provides unit recreation, library books, sports programs, and rest areas for brigade-sized and larger units.

HUMAN RESOURCES PLANNING AND STAFF OPERATIONS

10-13. Human resources planning and staff operations are the means by which the human resources provider envisions a desired human resources end state in support of the operational commander's mission requirements. Human resources planning addresses the effective ways of achieving success, communicates to subordinate human resources providers and human resources unit leaders the intent, expected requirements, and outcomes to be achieved, and provides the support OPLANs, OPORDs, or Planning Annex.

FINANCIAL MANAGEMENT

10-14. The financial management mission is to analyze the commander's tasks and priorities to ensure that proper financial resources are available to accomplish the mission and to provide recommendations to the commander on the best allocation of scarce resources. Financial management provides the capability for finance and RM operations across the theater to include all unified operations.

10-15. Financial management is comprised of two core functions: finance operations and RM. These two processes are similar and mutually supporting in organizational structure and focus.

FINANCE OPERATIONS

10-16. Finance operations must be responsive to the demands of the unit commanders at all levels, requiring financial management leaders to anticipate and initiate the finance support needed. Finance operations support is not SF-specific; USASOC has no organic financial management units and relies solely on conventional forces support. Finance operations support the sustainment of Army, joint, and multinational operations through the execution of key finance operations tasks:
- Provide timely commercial vendor and contractual payments, and various pay and disbursing services.
- Oversee and manage the Army's Banking Program.
- Implement financial management policies and guidance prescribed by the Office of the Under Secretary of Defense (Comptroller) and national financial management providers; for example, the U.S. Treasury, Defense Finance and Accounting Service, and Federal Reserve Bank.

10-17. The success of all operations depends on the support provided to the sustainment system and to contingency contracting efforts. By coordinating with the contracting officer and the staff judge advocate regarding local business practices, financial managers greatly reduce the probability of improper or illegal payments. Procurement support includes two areas: contracting support and commercial vendor services support.

Contracting Support

10-18. Contracting support involves payment to vendors for goods and services. This includes all classes of supply, laundry operations, bath operations, transportation, and maintenance. Financial managers are crucial to successful contracting operations.

Commercial Vendor Services Support

10-19. Commercial vendor services support provides for the immediate needs of the force. These are needs the standard logistics systems cannot support. This usually includes payments of cash (U.S. or local currency). Cash payments are usually for day laborers, Class I supplements (not otherwise on contract), and the purchase of construction material not readily available through the contract or supply system.

RESOURCE MANAGEMENT

10-20. The RM section is the commander's leading representative responsible for financial management support. RM advises the appropriate allocation and use of scarce resources, to include funding, in the accomplishment of the commander's assigned missions. RM personnel assist commanders by providing a critical capability, which matches legal and appropriate sources of funds with thoroughly vetted and valid requirements. Funding support provides flexibility through nonlethal methods to augment, and in some cases, lead the effort in obtaining the effects the commander is trying to achieve.

10-21. The RM mission is to analyze resource requirements ensuring commanders are aware of existing resource implications in order for them to make resource-informed decisions, and then obtain the necessary funding that allows the SF commander to accomplish the overall unit mission. Key RM tasks include the following:

- Providing advice and recommendations to the commander.
- Identifying sources of funds.
- Forecasting, capturing, analyzing, and managing costs.
- Acquiring funds.
- Distributing and controlling funds.
- Tracking costs and obligations.
- Establishing and managing reimbursement processes.
- Establishing and managing the Army Managers' Internal Control Program.

10-22. RM personnel also provide a variety of organic support to commanders for overseas contingency operations, Joint Chiefs of Staff exercises, counternarcotics training, and JCET events.

10-23. Regardless of the scale or scope of sustainment operations, finance and RM operations play a key role in providing responsive agile support to deployed forces across the spectrum of conflict. Each of these operations must be fully integrated and synchronized with all other facets of sustainment operations in order to effectively and efficiently sustain the force (FM 1-06 provides additional information). ARSOF units have some unique funding authorizations and their use must be clearly understood.

Operational Funds Support

10-24. RM organic support includes use of OPFUNDs. USASOC OPFUNDs are governed by USASOC Policy Number 32-09. OPFUNDs can be requested before, during, or after deployments and authorized training events. Units must work with their RM office to request an OPFUND. In general, the commander appoints a pay agent on an additional duty appointment order. This appointment authorizes the pay agent to disburse public currency IAW the special instructions stated in the appointment and the written instructions provided by the financial management commander. The field ordering officer, whom the pay agent supports, receives separate instructions from contracting officials. Field ordering officers and pay agents train and work as a team; the pay agent should participate with their finance operations in training and vice versa. The pay agent or finance operations may be held personally liable for any payment not IAW the appointment orders or prescribed instructions. The pay agent cannot simultaneously serve as either a certifying officer or finance operations. The pay agent uses an official credit or debit card to make payments whenever possible.

10-25. When it is not possible to use an official credit or debit card to make payments, the pay agent takes the following actions:

- Reviews all Standard Forms 44 prepared by the ordering officer.
- Disburses currency for the goods or services as stated on the Standard Form 44, but only after this form has been approved by a field ordering officer.
- Pays for purchases not to exceed established limits. (An agent may not split purchases between two or more vouchers to circumvent the established limit.)
- Clears his account with the disbursing officer that advanced the funds.

Other Funding Support

10-26. Funding support is a complex endeavor and requires RM personnel to leverage multiple appropriations. Some of these appropriations are initially provided for peacetime support, along with appropriations that are newly created by Congress specifically for an operation. Commanders and RM

Personnel Services Support

personnel need a thorough understanding of the statutes and regulations that govern the use of appropriated and nonappropriated funding. RM personnel must work closely with the fiscal lawyer to ensure compliance with fiscal requirements established by law. The following discussion highlights the basic appropriations that fund SOF. Multitudes of funding options are available and may include funding sources from other U.S. agencies (for example, intelligence funding, counterdrug funding, and DOS funding).

10-27. Funding authorities the financial management personnel may leverage before, during, and after contingency operations include the following:

- *Operations and Maintenance, Army (OMA) or MFP-2.* SF units receive some direct funding (MFP-2) from Army for some Army-common requirements. SF units use MFP-2 to pay for the day-to-day expenses in garrison, and during exercises, deployments, and military operations. There are threshold dollar limitations for certain types of expenditures, such as purchases of major end items of equipment and construction of permanent facilities. OMA is typically a one-year appropriation and must be obligated in that fiscal year (01 October to 30 September).
- *Operations and Maintenance, Defense (OMD) or MFP-11.* SF units use MFP-11 for training, equipping, and employing SF with SO-peculiar equipment, materials, supplies, and services. MFP-11 is for SOF-unique requirements only. The same dollar limitations apply to MFP-11 as MFP-2 funds.
- *Military Personnel, Army (MPA).* MPA funding is used for pay, allowances, individual clothing, subsistence, interest on deposits, gratuities, and permanent change of station travel (including all expenses for organizational movements) for members of the Regular Army and mobilized Reserve and National Guard Soldiers. MPA funding is generally available for one fiscal year and is centrally managed and funded. Since MPA funding is centrally managed, personnel should plan in advance for the use of MPA funding to ensure receipt in time to satisfy the requirement.
- *Procurement.* While OMA funds day-to-day operations, procurement is typically used for centrally managed items or systems that are considered investment items. These items require the use of procurement funds regardless of cost (or the cost of individual components). Such items can include large pieces of equipment or systems that exceed the expense investment threshold.
- *Research, development, test, and evaluation (RDT&E).* RDT&E funds provide for the development, engineering, design, purchase, fabrication, or modification of end items, weapons, equipment, or materials. This is not an appropriation normally used in the theater by deployed units unless involved in the research, development, acquisition, and testing process. RDT&E funding is available for two years.
- *Military construction.* Military construction provides for the acquisition of land and construction of buildings for which authorizing legislation is required.

10-28. In addition to the funding support provided by the sources listed above, the commander and RM personnel may leverage the additional funding sources listed in the following paragraphs in support of the mission.

Commanders' Emergency Response Program

10-29. The purpose of the CERP is to enable military commanders to respond to urgent humanitarian relief and reconstruction requirements within their AORs by carrying out programs that will immediately assist the local populace. The program is designed to allow commanders down to the ARSOF battalion level to have the ability to make an immediate, positive impact in their AOs/AORs. This authority is authorized through the enactment of annual authorization/appropriations acts and is not codified in law. DOD Financial Management Regulation 7000.14-R, Volume 12, *Special Accounts Funds and Programs*, Chapter 27 (Commanders' Emergency Response Program), provides implementing policy and guidance for the use of the CERP. The guidance primarily assigns administration responsibilities, defines proper CERP projects, and specifies accountability procedures. This guidance is mandatory reading for anyone intending to use CERP funds.

Chapter 10

Department of Defense Rewards Program

10-30. The DOD Rewards Program is not an intelligence program and is not intended to replace existing programs. DOD Financial Management Regulation 7000.14-R, Volume 12, Chapter 17 (DOD Rewards Program), provides overall policy and guidance for the implementation of the DOD Rewards Program.

10-31. Section 127b, Title 10, United States Code, *Assistance in Combating Terrorism: Rewards*, authorizes the DOD to pay rewards to persons for providing U.S. Government personnel or government personnel of multinational forces participating in a multinational operation with U.S. armed forces with information or nonlethal assistance that is beneficial to—

- An operation or activity of the armed forces or of multinational forces participating in a multinational operation with multinational forces conducted outside of the United States against international terrorism.
- Personnel protection of the armed forces or multinational forces participating in a combined operation with U.S. armed forces.

10-32. This authority is useful to encourage the local citizens of foreign countries to provide information and other assistance, including the delivery of dangerous personnel and weapons, to U.S. Government personnel or government personnel of multinational forces. GCCs provide additional policy guidance for this program within their respective AOR. U.S. or multinational units pay rewards for information helpful to the multinational forces and are not limited only to information leading to the capture of a high-value individual or seizure of weapons.

Emergency and Extraordinary Expense Authority

10-33. Emergency and extraordinary expense authority is found in Section 127, Title 10, United States Code, *Emergency and Extraordinary Expenses*. This provides the Secretary of Defense and Service secretary's authority to expend operations and maintenance (O&M) funds without regard to contracting and purpose limitations. This authority is provided annually in the O&M appropriations. Emergency and extraordinary expense funds are those that may be used to support certain unique requirements of operations. USSOCOM regulations cover these funds and define the types of acceptable expenditures. Very small amounts of this authority exist. The GCC can request the Service component to provide emergency and extraordinary expense funds. This authority does not provide cash or foreign currency to conduct an activity; rather, it provides the capability to obligate O&M funds for an activity normally not authorized for O&M funding. The USSOCOM receives emergency and extraordinary expense authority to support Official Representation Funds and Confidential Military Purpose funds:

- *Official Representation Funds*. These funds are used by high-level commanders (usually division commander and above) to uphold the standing and prestige of the United States by extending official courtesies to certain officials and dignitaries of the United States and foreign countries. Used correctly, Official Representation Funds are very helpful in building relationships in contingency operations. DOD Instruction 7250.13, *Use of Appropriated Funds for Official Representation Purposes*, should be referenced regarding proper obligation and expenditure of these funds.
- *Confidential Military Purpose funds*. These funds are used for operational preparation of the environment, to include advanced special operations.

Combatant Commander Initiative Fund

10-34. The Combatant Commander Initiative Fund provides a means for GCCs to react to unexpected contingencies and opportunities. Funds may be used for—

- Mission command.
- Joint exercises.
- Humanitarian and civic assistance.
- Military education and training to military and related civilian personnel of foreign countries.

- Personnel expenses of defense personnel participating in bilateral or regional cooperation programs and contingencies.
- Selected operations.

Section 1206 Authority

10-35. NDAA Section 1206 authority is used to conduct or support programs globally that build the capacity of a foreign country's military and maritime security forces. Types of equipment provided by this funding include—
- Radios and telecommunication systems.
- Surveillance and reconnaissance systems.
- Trucks, ambulances, boats, and other vehicles.
- Small arms and rifles.
- Night vision goggles and sights.
- Clothing.

Section 1208 Authority

10-36. NDAA Section 1208 authorizes assistance to foreign forces, irregular forces, groups, or individuals supporting U.S. counterterrorism military operations. Section 1208 also authorizes expenditure of funds to support—
- Operational preparation of the environment.
- Advanced special operations.
- Advanced force operations.
- Direct action in support of ongoing operations.

10-37. Lastly, Section 1208 authorizes the DOD to reimburse foreign forces, groups, or individuals supporting or facilitating ongoing counterterrorism military operations by SOF.

Sensitive Mission Fund

10-38. Sensitive Mission Funds are used when the mission requires the ability to conceal the payee, the payer, or the purpose of the payment using unique financial and accounting procedures.

Memorandum of Agreement

10-39. Memorandums of agreement are agreements between countries or eligible organizations that delineate responsibilities among the participants. Among these responsibilities are the participants' financial liabilities for support. These agreements define the specific mechanisms required for reimbursement of costs. Memorandums of agreement must be signed by the USASOC chief of staff and be based on specific legal authority and negotiated IAW proper procedures.

LEGAL SUPPORT

10-40. Members of The Judge Advocate General's Corps provide proactive legal support on all issues affecting the Army and the joint force and deliver quality legal services to Soldiers, retirees, and their families. Legal support centers on six core disciplines: military justice, international and operational law, administrative and civil law, contract and fiscal law, claims, and legal assistance. Each discipline is described below. (FM 1-04 provides additional information.)

MILITARY JUSTICE

10-41. Military justice is the administration of the Uniform Code of Military Justice. The Judge Advocate General is responsible for the overall supervision and administration of military justice within the Army. Commanders are responsible for the administration of military justice in their units and must communicate directly with their servicing staff judge advocates about military justice matters (AR 27-10, *Military Justice*).

Chapter 10

INTERNATIONAL AND OPERATIONAL LAW

10-42. During the conduct of operational missions, the command judge advocate section's involvement is crucial in articulating clear rules of engagement and use of force for the mission, as well as providing necessary guidance to ensure that the mission is conducted IAW the applicable international and domestic laws. Furthermore, the command judge advocate section reviews unit concept of operations to ensure that the mission or training conducted is within the legal authorities of the approving commander. When operating in a foreign country, SF units should seek assistance from the command judge advocate section regarding their legal status, such as the status-of-forces agreement, within that country. The command judge advocate section is tasked to provide mandated laws of armed conflict and human rights training before each deployment.

ADMINISTRATIVE AND CIVIL LAW

10-43. All investigation requirements, such as AR 15-6, *Procedures for Investigating Officers and Boards of Officers*; Commander's Inquiry; Line-of-Duty; Financial Liability Investigations and Property Loss; or Uniform Code of Military Justice Article 32, *Investigation*, require the assistance of the command judge advocate section. The command judge advocate section assists the commander and staff in setting the scope of the investigation, briefs the investigating officers how to conduct the investigation, and conducts legal reviews to ensure that the investigation is legally sufficient. During deployment, the command judge advocate section advises the commander on investigation requirements set in the AOR. In addition to investigations, the command judge advocate section provides ethics training and ethics guidance to ensure that SF personnel abide by the DOD ethics regulation. Because of the punitive nature of the DOD ethics regulation, it is crucial that all SF personnel are familiar with and abide by the DOD ethics regulation.

CONTRACT AND FISCAL LAW

10-44. Contract law is the application of domestic and international law to the acquisition of goods, services, and construction. The practice of contract law includes battlefield acquisition, contingency contracting, bid protests and contract dispute litigation, procurement fraud oversight, commercial activities, and ACSAs. The staff judge advocate's contract law responsibilities include furnishing legal advice and assistance to procurement officials during all phases of the contracting process, overseeing an effective procurement fraud abatement program, and providing legal advice to the command concerning battlefield acquisition, contingency contracting, use of logistics civilian augmentation program, ACSAs, the commercial activities program, and overseas real estate and construction.

10-45. Because of the restrictive nature of fiscal law, it is crucial for SF units to obtain legal guidance and review prior to making purchases or contracting. At the SFODA/SFODB levels, any questions regarding the use of OPFUNDs should be reviewed by the command judge advocate section to ensure that there is proper justification and authority to make such a transaction. Furthermore, any drafting of statements of work or performance work statements for a contract should get reviewed by the command judge advocate to ensure proper language is used.

CLAIMS

10-46. IAW AR 27-20, *Claims*, any claims against the U.S. Government because of unit actions should be referred to the command judge advocate for assistance.

LEGAL ASSISTANCE

10-47. Legal assistance is the provision of personal civil legal services to Soldiers, their family members, and other eligible personnel. The command judge advocate section is able to provide limited legal assistance, such as drafting wills and powers of attorney. From an operational standpoint, the provision of legal services at the earliest possible time is critical to ensure the readiness of individual Soldiers and the force as a whole.

RELIGIOUS SUPPORT

10-48. Religious support facilitates the free exercise of religion, provides religious activities, and advises commands on matters of morals and morale. Chaplains and chaplain assistants functioning as unit ministry teams perform and provide religious support in the Army to ensure the free exercise of religion (FM 1-05 has additional information). The Chaplain is responsible for providing direct religious support in three ways: nurture the living, care for the wounded, and honor the dead with memorial ceremonies and services. His missions are to advise the commander on all matters of religion, ethics, morals, and morale within the unit.

BAND SUPPORT

10-49. USASOC has no organic band units and relies solely on conventional forces support (FM 1-0 and ATTP 1-19 provide additional information). Army bands provide critical support to the force by tailoring music support throughout military operations. Music instills in Soldiers the will to fight and win, foster the support of U.S. citizens, and promote America's interests at home and abroad.

This page intentionally left blank.

Chapter 11
Contracting and Host-Nation Support

ARSOF units have no organic operational contracting capability and are reliant upon contracting support from the theater GCC, Service components, and USASOC or USSOCOM contracting activities. The key to successful contingency contracting execution in support of SF is the preplanning and early identification of the goods and services that are required. SO may have special mission requirements coupled with the diversity and complexity of assigned missions that may require an integrated approach from different contracting activities. This is especially important when working with HN vendors and indigenous sources of goods and services.

TYPES OF CONTRACTOR SUPPORT

11-1. There are numerous types of contractor support. Program, project, and product managers use systems support contractors to support the Army's acquisition system. Numerous other types of contracts and contractor support are available to support the ARSOF commander's requirements. Logisticians of each command must be familiar with the various types of contractor support available in order to better advise their commands.

SYSTEMS SUPPORT CONTRACTORS

11-2. Systems support contractors support many different Army materiel systems under prearranged contracts awarded by the Assistant Secretary of the Army for Acquisition, Logistics, and Technology (ASA[ALT]) program executive officer, program manager offices and the USAMC's Simulations, Training, and Instrumentation Command. Supported systems include, but are not limited to, newly- or partially-fielded vehicles, weapon systems, aircraft, and communications and computer infrastructure, such as the Army Battle Command Systems and Standard Army Management Information Systems and communications equipment. Systems support contractors, made up mostly of U.S. citizens, provide support in garrison and may deploy with the force to support training, contingency operations, and crisis response. They may provide either temporary support during the initial fielding of a system, called interim contracted support, or long-term support for selected materiel systems, often referred to as contractor logistics support.

EXTERNAL SUPPORT CONTRACTORS

11-3. External support contractors provide a variety of logistics and sustainment support to deployed forces. External support contracts are let by contracting officers from support organizations such as USAMC and the U.S. Army Corps of Engineers. They may be prearranged contracts or contracts awarded during the contingency itself to support the mission and may include a mix of U.S. citizens, third-country nationals, and local national subcontractor employees. External support contracts include the logistics civilian augmentation program, commercial sealift support, and leased real estate operational contracting support.

Generating Force Contracting

11-4. The USASOC Deputy Chief of Staff, Acquisition and Contracting (DCSAC) provides SO-unique contracted procurement capacity to USASOC units during force generation. DCSAC contracting authority is derived from USSOCOM. Primarily, DCSAC obligates MFP-11 funding through contracts written by

Chapter 11

their office in USASOC HQ. Installation contracting offices provide O&M contracted procurement capacity to SO units during force generation. Installation contracting offices' contracting authority is derived from the Mission and Installation Contracting Command. Primarily, installation contracting offices obligate Army O&M funding through contracts written at the installation they support. The DCSAC possesses the capacity to manage a field ordering officer program during force generation to support ARSOF training at Fort Bragg, North Carolina.

Operational and Exercise Contracting

11-5. The Lead Service for Contracting (LSC) supporting a GCC typically provides contracting support to SF during operations (deployments) and exercises (Joint Chiefs of Staff or JCET). The LSC varies depending on the GCC and by country within some GCCs. Expeditionary Contracting Command (ECC) contracting support brigades collocate with and support ASCCs worldwide. Depending on the theater, the ECC contracting support brigade may be the LSC.

11-6. The exception to the LSC rule is SOCCENT Contracting. SOCCENT is the only TSOC with expeditionary contracting capacity derived from USSOCOM. SOCCENT Contracting is responsible for SOF-unique contracting and contracting support where no other contracting capacity exists in the USCENTCOM AOR. As a potential operation or exercise force provider to GCCs, the ECC may provide CCOs to support an LSC contracting office or SOCCENT Contracting. When the ECC is not the LSC in the area and a CCO is required, support must be requested through either the Joint Capability Requirements Manager (JCRM) or the Joint Training Information Management System. The JCRM is used for operational support and the Joint Training Information Management System is used for training support.

11-7. Contingency contracting support is provided to units with validated requests based on CCO availability across the Services (among other factors). A request through either the JCRM or the Joint Training Information Management System does not guarantee ECC CCO support. LSCs and SOCCENT Contracting have the capability to manage field ordering officer programs during operations and exercises. Installation contracting offices have the capability to manage field ordering officer programs to support short-duration (60 days) exercises when SF units deploy from home station and return to home station. Installation contracting offices perform this function only when no other contracting support is required during the exercise.

PLANNING CONTRACT SUPPORT

11-8. The ECC functions as the operational and exercise contracting support planner and the predeployment contingency contracting support trainers to SO units. The ECC is the SO units advocate and liaison to bring about effective and efficient operational and exercise contracting support through coordination with combatant command LSCs and SOCCENT Contracting. Contingency contracting unit training for contracting officer representative and field ordering officer is available during predeployment. ECC contingency contracting teams collocated with SFG(A)s perform duty in installation contracting offices. SF units requiring planning support, operational and exercise contracting support coordination, or contingency contracting unit training should contact ECC contingency contracting teams.

KEY CONTRACTING OFFICIALS

11-9. The following sections describe in general the roles, responsibilities, and authorities of key contracting officials who participate in, manage, and oversee the contracting within the SO community.

HEAD OF CONTRACTING AUTHORITY

11-10. The head of contracting authority is a general officer, usually the senior commander in the theater or a deputy designated by that commander, who provides overall guidance throughout the campaign or operation. The head of contracting authority serves as the approving authority for contracting as stipulated in regulatory contracting guidance. The head of contracting authority appoints the principal assistant responsible for contracting (PARC). All Army contracting authority in a theater flows from the head of contracting authority to the PARC.

Principal Assistant Responsible for Contracting

11-11. The PARC, a special staff officer, is the ASCC or mission commander's senior Army acquisition advisor responsible for planning and managing all Army contracting functions within the theater. All Army contracting authority in a theater flows from the head of contracting authority to the Army's PARC. All Army contracting personnel within the theater operate under the procurement authority of the PARC. The PARC's functional control of contracting requires all contracting personnel from any Army agency or supporting command to coordinate their activities with the PARC. Functional control is normally accomplished through the acquisition review board process and follows the PARC's contracting support plan or acquisition instructions when procuring all goods or services within the theater. In a joint environment, the PARC may be the designated executive agency for theater contracting, with responsibility to coordinate all DOD contracting activities.

Contracting Officer

11-12. The contracting officer is an official with the legal authority to enter into, administer, and terminate contracts. A contracting officer is appointed in writing through a warrant (Standard Form 1402 [Certificate of Appointment]) by a head of contracting authority or a PARC. Only duly warranted contracting officers, appointed in writing, or their designated representatives are authorized to obligate the U.S. Government. Active Army and USAR military personnel, as well as DA civilian personnel, may serve as contracting officers supporting deployed Army forces.

Contracting Officer Representative

11-13. A contracting officer representative is an individual appointed in writing by a contracting officer to act as the eyes and ears of the contracting officer. This individual is not normally a member of the contracting organization, but most often comes from the requesting unit or activity. The contracting officer assigns the contracting officer representative specific responsibilities, with limitations of authority, in writing. The contracting officer representative represents the contracting officer only to the extent documented in the written appointment.

Field Ordering Officer

11-14. A field ordering officer is an individual who has written authorization from a warranted contracting officer to sign a contract instrument for micro-purchases. Neither PBOs nor paying agents may be ordering officers.

Paying Agent

11-15. A paying agent is also known as a Class A agent. The purpose of a paying agent is to make specific payments. Paying agents are appointed to the position of paying agent under the exclusive supervision of the disbursing officer in all matters concerning custody and disposition of funds advanced to them. Paying agents will comply with all instructions and regulations pertaining to their paying agent duties as issued by the disbursing officer from the finance detachment. Funds advanced to a paying agent are held at personal risk by the paying agent and must be accounted for to the disbursing officer immediately upon completion of the transaction for which advanced.

Requiring Activity

11-16. A requiring unit or activity is that organization or agency that identifies a specific sustainment requirement through its planning process to support the mission. All requiring units or activities are responsible to provide contracting and contractor oversight in their respective AO. This oversight is accomplished through appointed contracting officer representatives, to include submitting contractor accountability and visibility reports, as required. Requiring units can either be a tactical- or operational-level unit in the AO or a sustainment organization, such as an ASA(ALT) project executive office/project management or USAMC, which has identified a support requirement that affects forces in the field. This organization identifies the specific support requirements. If it is determined that the requirement is best satisfied by contractor support, this organization prepares the required statement of work that supports the contracting process. The requiring unit or activity may not be the organization actually receiving the contractor support. These units are simply referred to as the supported unit.

United States Special Operations Command Contracting Activity Organization

11-17. The USSOCOM PARC is the procurement authority that gives all ARSOF contracting officers their procurement authority. The USSOCOM PARC is based at USSOCOM HQ and operates under the Special Operations Federal Acquisition Regulation.

Contractor Functions on the Battlefield

11-18. Contractor support is categorized by the type of support that is provided on the battlefield and, more importantly, by what type of contracting organization has contracting authority over them. Battlefield contractors are generally referred to as theater support contractors, external support contractors, or systems support contractors. Commanders and planners must be aware that a requirement for a particular system or capability may result in the introduction of these types of contractors into the campaign or operational plan. Contractor management and planning is often significantly different depending on the type of contractor support provided.

MANAGING CONTRACTORS ON THE BATTLEFIELD

11-19. Effective contractor management on the battlefield is essential to ensure that contractor-provided support is properly orchestrated and synchronized with the overall operation support plan. Commanders must ensure that contractor employees are properly accounted for, protected, and supported. Additionally, adequate contractor-employee accountability and contractor visibility in the theater is necessary to establish positive control, to perform initial reception and integration, to provide necessary support, and to establish and manage their location and movement on the battlefield. Further information is provided in ATTP 4-10; the Defense Acquisition Guidebook; and AR 715-9, *Operational Contract Support Planning and Management*.

HOST-NATION SUPPORT AND ACQUISITION AND CROSS-SERVICING AGREEMENTS

11-20. HNS is civil and military assistance rendered by a nation to foreign forces within its territory during peacetime, crises or emergencies, or war based on agreements mutually concluded between nations. Many HNS agreements have already been negotiated between NATO nations. Potential HNS agreements may address labor support arrangements for port and terminal operations, using available transportation assets in country, using bulk petroleum distribution and storage facilities, possible supply of Class III (Bulk) and Class IV items, and developing and using field services. The United States initiates and continually evaluates agreements with multinational partners for improvement. They should be specifically worded to enable planners to adjust for specified requirements. Additionally, the commander should assess the risk associated with using HNS, considering operational area security and operational requirements. Planners must realize that HNS may not meet U.S. standards.

11-21. Under ACSA authority (Section 2341, Title 10, United States Code, *Authority to Acquire Logistic Support, Supplies, and Services for Elements of the Armed Forces Deployed Outside the United States*, and Section 2342, Title 10, United States Code, *Cross-Servicing Agreements*), the Secretary of Defense can enter into agreements for the acquisition or cross-service of logistics support, supplies, and services on a reimbursable replacement-in-kind, or exchange-for-equal-value basis. These agreements can be with eligible nations and international organizations of which the United States is a member. An ACSA is a broad overall agreement, which is generally supplemented with an implementing agreement. The implementing agreement contains points of contact and specific details of the transaction and payment procedures for orders for logistics support. Neither party is obligated until the order is accepted.

11-22. Under these agreements, common logistics support includes food, billeting, transportation (including airlift), POL, clothing, communications services, medical services, ammunition, base operations, storage services, use of facilities, training services, spare parts and components, repair and maintenance services, calibration services, and port services. Specific items that may not be acquired or transferred under ACSA authority include guided missiles; naval mines and torpedoes; nuclear ammunition and included items, such as warheads, warhead sections, projectiles, demolition munitions, and training ammunition; cartridge and propellant-actuated devices; chaff and chaff dispensers; guidance kits for bombs or other ammunition; and chemical ammunition (other than riot control agents). General purpose vehicles and other items of nonlethal military equipment not designated as significant military equipment on the United States Munitions List promulgated pursuant to Section 2778, Title 22, United States Code, *Control of Arms Exports and Imports*, may be leased or loaned for temporary use. Specific questions on the applicability of certain items should be referred to the combatant command's legal office for review and approval.

This page intentionally left blank.

Appendix A
Logistics Planning Checklist

The purpose of the logistics planning checklist (Figure A-1, pages A-1 through A-8) is to provide a tool for logistics planners to use in support of ARSOF. It is not all-inclusive; however, it serves as a point of departure for the planning of ARSOF support and sustainment. It is extremely detailed and may be used to check the thoroughness of any SO logistics support plan. The checklist separates a support plan into its fundamental parts and presents questions for evaluating the content.

REFERENCES
List doctrinal, policy, and procedural publications appropriate to the level at which the plan is prepared:
- Do any CONPLANs apply?
- Are the necessary maps listed and available?

PURPOSE
Provide a concise statement of the purpose for which the logistics support plan is prepared.

GENERAL
Provide a summary of the requirements, taskings, and concept of operations that the logistics planning supports:
- Are the objectives specified?

ASSUMPTIONS
List the assumptions upon which the concept of operations and logistics support are based.

RESPONSIBILITIES
Are responsibilities for support clearly stated for the following?
- Joint staff.
- USSOCOM.
- United States Transportation Command (USTRANSCOM).
- Other military Services.
- Combatant commands and their component commands.
- Theater SO commands.
- Defense Security Assistance Agency.
- National Geospatial-Intelligence Agency.
- DOS and U.S. Embassies.
- Security assistance organizations.
- Liaison offices.
- Defense Logistics Agency.
- Army and Air Force exchange service.
- Units or elements providing logistics support to ARSOF components.

Figure A-1. Logistics planning checklist

Appendix A

CONCEPT OF LOGISTICS SUPPORT
Analyze the sufficiency of the concept of logistics support: <ul><li>How will supply, maintenance, transportation, and field service support be provided?</li><li>Which logistics elements will provide the support? Are the forces that provide the support adequate?</li><li>Does the planned support complement the tactical plan? Is it adequate and feasible?</li><li>How will the terrain and enemy intelligence impact on logistics support?</li><li>Has the deployment flow been properly analyzed to determine the time-phasing for introduction of logistics elements to support the combat forces?</li><li>Is HNS available and what are the subsequent risks?</li><li>What support will SO-peculiar equipment, materials, supplies, and services provide?</li><li>What are the validation procedures for SO-peculiar equipment, materials, supplies, and services?</li></ul>
SUPPLY
Validate whether adequate supply support has been planned in the following areas.
General
Consider basic supply procedures: <ul><li>Are the supply system and procedural guidance provided?</li><li>Is the flow of requisitions and materiel described?</li><li>Is a project code required and identified?</li><li>Is a temporary force activity designator upgrade required?</li><li>Are in-country DOD activity address codes required?</li><li>Are ALOC procedures described?</li><li>Is the number of days of supplies required to accompany troops identified?</li><li>Are provisions made for contracting and local purchase support?</li><li>Are the stockage objectives by classes of supply specified?</li><li>Will automated or nonautomated procedures be used?</li><li>Will automated systems of supported units and task-organized logistics units interface?</li><li>Have the inter-Service support requirements been identified and common-, cross-, and joint-servicing arrangements coordinated for support of ARSOF?</li><li>What HNS will be provided?</li><li>What theater of operations support is required?</li><li>Are retrograde procedures specified for excess and unserviceable items?</li><li>What are the provisions for emergency resupply?</li><li>Have initial preplanned supply support and emergency support packages been considered?</li><li>Is the communications capability provided and compatible with the automated systems being deployed?</li><li>Are changes to the DOD activity address file required, such as "ship-to address"?</li><li>Are some supply support activities to be designated as ALOCs?</li><li>Are procedures described for cancellation or diversion of materiel in-process or in-transit at the termination of the operation or exercise?</li><li>Are there provisions for logistics support of displaced civilians, prisoners of war, and indigenous personnel?</li><li>Is there covered storage in the AO to protect supplies from the elements? If not, are shipments packed for outdoor storage?</li><li>Are MHE requirements provided?</li></ul>

Figure A-1. Logistics planning checklist (continued)

- Is sufficient rigging material and equipment available for airdrop operations?
- Is the defense automatic addressing system aware of the communications routing identifier and DOD activity address code for processing direct requisitions and direct supply status?
- What are the procedures for distributing maps?

Class I
Consider subsistence issues:
- Are mess facilities identified and adequate?
- Are the ration cycles described by phase? Is a ration cycle proposed?
- Are fresh eggs, fresh fruits and vegetables, fresh meats, juices, milk, and canned soft-drink supplements to the meal, ready to eat (MRE) and other field ration meals considered?
- Do local fresh fruits and vegetables meet U.S. standards?
- Have unitized operational rations been considered for ease of handling and accountability?
- Are cash meal-payment procedures established?
- What method of distribution will be used (unit distribution or supply point distribution)?
- Are bakery supplements to the field ration meals considered?
- Are veterinary personnel adequate for the subsistence support requirements?
- Are hospital rations required?
- Are chill, freeze, and refrigeration requirements for unit dining facilities and Class I supply points addressed?
- Are water support requirements satisfied?
- Are the sources of water fresh, brackish, or salty?
- Is the source of water from local systems surface or wells?
- What type of water purification unit is required?
- Are chillers required?
- What is the water-planning factor in gallon per man per day?
- What are the treatment, storage, distribution, and cooling requirements? Are they satisfied by deploying unit capability?
- What are the well-drilling requirements? Are there any existing wells? What is the quality of water from existing wells?
- Are potable ice considerations covered? What is the requirement planning factor? Have the AHS planners provided for certification of ice as potable?

Class II
Consider clothing, individual equipment, tools, and administrative supplies issues:
- Are requirements for individual clothing and mission-essential consumables addressed?
- What are the requirements for mission rehearsals and training?
- What are the provisions for replacing damaged personal clothing and chemical protective clothing?
- Which self-service supply center (SSSC) listing will be used as the basis for the Class II stockage?
- How will the logistics support element replenish organizational clothing and individual equipment and SSSC items?
- Do any of the following items require special considerations?
    - SO-peculiar materials.
    - Tentage and tentage repair kits.
    - Administrative and office supplies.
    - Folding cots.

Figure A-1. Logistics planning checklist (continued)

Appendix A

- Insect bars with mosquito netting.
- Banding material and tools.
- Water-purification chemicals and test kits.
- Insect repellent and sun screen.
- Field laundry and bath supplies and hospital laundry supplies.
- Dining facility supplies, including paper and plastic products.
- Trash disposal supplies.
- Vector control equipment and supplies.
- Latrine chemicals and supplies.
- Batteries.
- Cold weather clothing and equipment.
- Air conditioners or fans.

Class III

Consider POL requirements and support:
- Are service requirements by location for each type product established?
- Is the use of contractor-provided bulk fuels considered?
- Are ordering and accountable officer requirements addressed?
- Are existing pipeline distribution systems available? What are the pipeline and storage capabilities?
- Are remote refueling sites or FARPs required? What capabilities are required?
- Are inter-Service support billing and reimbursement procedures specified?
- Are POL quality surveillance procedures specified? Are required test kits on hand?
- Is a petroleum laboratory available?
- Are additives required for commercial fuels? Who will provide them?
- Are any unique package product requirements addressed?
- Are industrial gases addressed?

Class IV

Consider construction materials requirements:
- Are unique requirements for construction, security, and rehearsal materials addressed?
- Is in-country procurement considered?
- Have Class IV data sources been queried on preexisting databases describing locally available construction materials?
- Are basic loads to be deployed?
- Will the use of pre-positioned material stocks be permitted?

Class V

Consider ammunition requirements:
- Are unit basic loads to be deployed?
- Is the logistics support structure prescribed?
- Are explosive ordnance disposal support requirements and procedures addressed?
- Are SO-peculiar ammunition requirements addressed?
- Have the storage, handling, shipping, security, and safety requirements been reviewed and addressed in the planning?
- Are requirements identified by category of munitions?
- Are sustaining rates of munitions addressed?
- Are special permits needed? Who issues them?

Figure A-1. Logistics planning checklist (continued)

Class VI
Consider issues of personal demand items:
- Are the deploying personnel provided guidance on personal demand items?
- Are sundry packs available?
- Is indirect or direct exchange support considered?
- If exchange support is required—
  - Has the Army and Air Force Exchange Service HQ been notified?
  - Have the exchange staffing, stock assortment, security, facility, transportation, and communications requirements been identified and coordinated?
  - Is finance support for the exchange identified?
  - Has the policy on rationing and check-cashing been determined?

Class VII
Consider major end-item requirements:
- Are SO-peculiar equipment requirements identified and validation procedures established?
- Does the plan specify the equipment-fill level for deploying units?
- Are equipment redistribution (cross-leveling) requirements specified?
- Are replacement actions for salvage equipment specified?
- Are operational readiness float requirements addressed?

Class VIII
Consider medical supply and material requirements:
- Are medical supply procedures prescribed?
- Does this portion of the logistics support plan complement the medical support plan?
- Are medical resupply procedures established?
- If applicable, are policies provided for the medical treatment of non-U.S. personnel?
- Are special medical equipment and supply requirements identified based on medical mission and the AO?
- Are memorandums of understanding established with medical logistics providers to ensure these medical supplies are stored, maintained, and ready to meet all operational contingencies?
- Are special storage requirements satisfied?
- Is the disposal of salvage medical supplies addressed?
- Are medical oxygen and other medical gases requirements (such as anesthesia identified and resupply procedures) established?
- Is local purchasing an option? Are procedures and guidelines established?
- Are procedures in place for identifying and replacing potency and dated materiel?

Class IX
Consider repair part requirements:
- Are SO-peculiar repair requirements specified?
- Are common repair parts requirements, including repairables, specified?
- Are cannibalization procedures addressed?
- Are requirements for nonexpendable components addressed?
- Is stockage of major assemblies addressed?
- Are there special storage requirements for such items as dry batteries, classified repair parts, and high-dollar materials?
- What are the procedures for disposing of hazardous materials, such as lithium batteries and radioactive residue?

Figure A-1. Logistics planning checklist (continued)

Appendix A

Class X
Consider the requirement for nonmilitary program materials:
- If Class X materials are required, does the plan describe the source?
- What is the source of funding for Class X supplies?

MAINTENANCE
Verify the validity of the support plan for maintenance:
- Does the plan describe how field, sustainment, and SO-peculiar equipment maintenance will be performed?
- Is missile maintenance support required and available?
- Does the plan address calibration requirements?
- Is maintenance exchange addressed?
- Have extreme weather aspects like heat, cold, humidity, and dust been considered?
- Are site security and storage requirements identified?
- Are special power requirements for maintenance facilities identified (for example, voltage, phase, frequency, stability, and anticipated load in kilowatts)?
- Are building suitability screening factors identified by type of maintenance facility (for example, minimum height and width for doors, floor load-bearing requirements, and environmental control necessities)?
- Are operational readiness floats addressed?
- How will repairs under warranty be performed in the AO?
- Is the evacuation of unserviceable repairable items addressed?
- What are the procedures for replacing maintenance tools and equipment?

TRANSPORTATION
Validate whether adequate transportation support has been planned in the following areas.

General
Consider basic transportation procedures and requirements:
- Is there a requirement for expedited cargo distribution to the AO?
- Are the transportation support systems for supply distribution and ALOC validation procedures outlined?
- What are MHE requirements?
- What is the availability of USTRANSCOM and Defense Intelligence Agency data analysis regarding the country's transportation infrastructure, including ports, airfields, roads, railroads, and inland waterways?
- Is a rail system available? What are the schedules and capability?
- Is the highway net described? What are the capabilities and limitations?
- What is the weather impact on ports, airfields, and highway nets?
- Are in-country highway, rail, air, and inland waterway mode requirements addressed?
- Are the transportation movement priority and transportation account codes provided? Are transportation funding procedures established?
- Has a dedicated in-country, intratheater, or intertheater movement system for personnel and high-priority cargo been established?
- Has coordination been made with USTRANSCOM for personnel and equipment movements?
- Has the use of foreign flag carriers been addressed?
- What agency will accept and coordinate administrative transportation requirements for SOF?
- What HNS is available?
- Are MEDEVAC requirements included in the planning?

Figure A-1. Logistics planning checklist (continued)

Airfields

Consider issues pertaining to airfield requirements:
- What airfields are available to support military operations?
- Is a coordinating HQ designated for all airlift support?
- Has support been planned for USAF mobile aeromedical staging facilities?
- What are the personnel and cargo reception capabilities of the aerial port of embarkation and aerial port of debarkation?
- What is the current usage of the airfield?
- What are the characteristics and capabilities of the roads that access the airfield?
- What contract civilian or HN personnel and equipment assets are available to assist at the aerial port of embarkation and aerial port of debarkation?
- Has an arrival/departure airfield control group (A/DACG) organization been designated? Have aerial port squadron and airlift control element requirements been identified?
- What airfield facilities are available for military use during A/DACG operations?
- What is the best source for additional information on the airfields?
- Have channel airlift requirements been specified?
- Have airbase defense requirements been properly addressed?

Supply Routes

Consider issues pertaining to the supply routes:
- What are the road movement and convoy restrictions?
- What routes are available to support military operations?
- What are the characteristics and capabilities of the routes available to support military operations?
- What are the dimensions and classifications of tunnels and bridges along the routes?
- What capabilities exist to repair damaged segments of routes?
- What segments of the routes are heavily used by the civilian populace?
- What are the most likely routes fleeing refugees would use?
- Are traffic-control measures in-place?
- What is the best source for additional information on the routes?

FIELD SERVICES

Consider issues pertaining to field services requirements:
- Are laundry, bath, clothing renovation, and latrine requirements addressed? Local sources?
- Are mortuary affairs capabilities adequate to support the anticipated requirements?
- Are procedures for salvage collection, evacuation, and disposal covered?
- Are base and post exchange services required and provided?
- Is fire protection provided for aviation, ammunition, and bases?
- Are procedures for waste disposal addressed?
- Are procedures specified and do units have the equipment necessary for cleaning of equipment for redeployment to meet customs and agriculture requirements to enter the CONUS?

MISCELLANEOUS

Consider miscellaneous issues and requirements:
- What are the billeting and support requirements at ISBs and SOTFs?
- Are HN military personnel with experience in U.S. military schools identified?
- Have arrangements been made with U.S. and HN customs and immigration?
- Are procedures for logistics reporting established?

Figure A-1. Logistics planning checklist (continued)

Appendix A

- Is delousing support required?
- Are isolation or rehearsal facilities required?
- What are the funding aspects of logistics support?
    - Are the costs for all requirements identified?
    - What is the account processing code?
    - Are SO-peculiar equipment resourcing procedures identified?
- What are the electrical power cycles of the country? Are transformers required?
- Are printing and duplicating requirements identified?
- Are the communications to support logistics operations included in the communications planning? Telephone?
- What are the requirements for aerial delivery, personal parachutes, and air items?
- Is a source of liquid oxygen required?
- What are the diving-support requirements?
- What are the administrative-use vehicle requirements?
- What are the audiovisual requirements?
- Are communication frequencies cleared with the HN government?
- Are there adequate provisions in the plan for contracting support?
    - Is an adequate number of contracting officers with the proper warrant provided?
    - Is adequate finance support available?
    - Is adequate legal support available?
    - Is adequate linguist support available?
    - Are there provisions in the plan for maneuver or war damage claims resulting from logistics operations?
- Are automated logistics systems procedures properly addressed?
    - Have backup master files been established and prepared for shipment separate from the primary master files?
    - Are maintainers, operators, and managers assigned and well-trained?
    - Have site selection and preparation for automated equipment considered accessibility, geographic, terrain, and security requirements?
    - Is there a continuity of operations plan?
    - Are sufficient copies of user manuals on hand and current?
    - Are sufficient repair parts available for the computer hardware, including generators and other subsystems?
    - Are there provisions for backup support for repair parts, hardware maintenance, and the receipt of software change packages?
    - Has telephone support been arranged?
    - Have details been worked out for transmission of documents to higher and lower echelons?
    - Will customer units require training? Are customer user manuals available for automated system support?
- Are operations security requirements integrated into logistics planning? Is the logistics signature minimized?
- Have security police requirements for SO bases, facilities, training areas, rehearsal sites, and storage sites been identified and resourced?

Figure A-1. Logistics planning checklist (continued)

Appendix B
Joint Operational Stocks

The JOS program is a joint, centrally managed, stored, and maintained stock of USSOCOM materiel. This program loans mission-critical and mission-essential equipment that directly supports TSOCs, components, and SOF units in the execution of training, contingency, and real-world missions. The JOS program is centrally managed within USSOCOM HQ. The Center for Special Operations (SCSO-J-3) validates and prioritizes loan requests; the Special Operations Acquisition and Logistics (SOAL) Center manages all logistics aspects of the program. The actual JOS inventory is maintained, stored, and issued from the SOFSA located in Lexington, Kentucky.

THE JOS CATALOG

B-1. The JOS Catalog contains a listing and a description of equipment and accessories contained in the JOS program, with information on policy and procedures. While this catalog is published and distributed in hard copy on an annual basis, it is also accessible online through the Special Operations Forces Sustainment, Asset Visibility, and Information Exchange (SSAVIE) home page. Updates are posted as they occur to the JOS Catalog online.

B-2. The JOS Catalog is only published once a year and initial distribution is made to units throughout the command via USSOCOM's Special Operations Forces Logistics Assistance Representatives. Additionally, copies will be sent to USSOCOM's Special Operations Forces Logistics Assistance Representatives for ease of access. Further, users may request the JOS Catalog by writing or calling the following:

Mailing Address:
Special Operations Forces Support Activity
5749 Briar Hill Road
M/S: 24 (SPO-JOS Project Manager)
Lexington, KY 40516

Phone Numbers:
Defense Switched Network (DSN): 745-3805
Commercial (Comm): (859) 293-3805

B-3. A hard copy of the JOS Catalog can be downloaded and printed off the JOS home page on the SSAVIE Web site. This copy is updated with new equipment throughout the year. To access the online JOS Catalog, users can log on to https://ssavie.sofsa.mil, and select the JOS button on the SSAVIE home page. However, users must first have a SSAVIE password to access the Web site. Passwords may be requested by clicking on the "SSAVIE ACCESS REQUEST FORM" button and completing the form that pops up.

TYPES OF JOINT OPERATIONAL STOCKS LOANS

B-4. The JOS Catalog provides various loan methods that units may use. There are new, extension, consolidation, and hand-off loan processes. Each unit or PBO should discuss which of the following types of loan to use when deciding about equipment purchases.

Appendix B

NEW LOAN

B-5. Units are directed to maintain control of JOS loans through unit PBOs or equipment custodians. New JOS loans will be processed through unit channels IAW procedures outlined in this catalog. While JOS loans typically cover a 90-day period (exclusive of the transportation pipeline), the length of the loan can be adjusted to correspond to the length of the deployment or exercise. However, loans will not generally be approved to exceed 6 months (180 days). The 180-day limitation is established to overcome some of the accountability problems associated with unit rotations and personnel turbulence.

LOAN EXTENSION

B-6. Units may request loan extensions when operations dictate and the loan cannot be returned as scheduled. Loan extension will be made for no more than 180 days each. Units will process extensions for loan in the same manner that they process new loan requests. Once processed, units will be required to sign updated JOS Loan Property Accountability Reports (hand receipts).

LOAN CONSOLIDATION

B-7. Units that have taken out multiple JOS loans may request that the loans be consolidated into a single loan to aid in accounting. Units need to be clear about which loans and what items of equipment are being consolidated.

HAND-OFF OR TRANSFER OF JOINT OPERATIONAL STOCKS LOAN

B-8. Hand-off or transfer of JOS equipment will be coordinated between PBOs and/or equipment custodians. Hand-off or transfer of JOS loaned equipment from one unit to another may occur because of operational considerations. When it occurs, the losing unit will obtain a signed hand receipt from the gaining unit for the equipment. A copy of this hand receipt will be sent along with the JOS request to transfer the equipment to the gaining unit. Responsibility for initiating the request for transfer of JOS loan rests with the losing unit. A new JOS loan will be created for the gaining unit. The gaining unit signs for the equipment under the new JOS loan. If all equipment is transferred from the old loan to the new loan, the loan is closed. If some equipment remains on the old JOS loan, the losing unit continues to be responsible for it.

Note: USSOCOM may, at any time, direct units to return JOS equipment to meet higher-priority missions or operational requirements.

PRIORITY OF JOINT OPERATIONAL STOCKS LOANS

B-9. JOS loans are normally prioritized in the order listed below. Exceptions are determined and approved by USSOCOM Center for Special Operations (SCSO-J-3).
- Priority 1: President-directed.
- Priority 2: Joint Chiefs of Staff-directed.
- Priority 3: Security assistance/humanitarian assistance missions (mobile training teams/ humanitarian demining operations).
- Priority 4: JCET events.
- Priority 5: Other exercises/unilateral training.

JOINT OPERATIONAL STOCKS LOAN APPLICATION PROCEDURE

B-10. Units/agencies requesting JOS equipment should first identify the equipment required (see annexes in the JOS Catalog) and then prepare a unit JOS loan request using the JOS Loan Worksheet contained in enclosures D-1 and D-2. The request may be submitted via letter, message, memorandum, or facsimile (Fax). However, these requests must be submitted through the requesting unit's chain of command for

Joint Operational Stocks

approval, or through the JOS Online Loan Processing System (component command approval is required to exercise this feature). SOF units under operational control of theater GCCs will submit their requests through their TSOC.

B-11. Requesting units must ensure that the mission category is identified, and that a short unclassified mission impact statement is provided in the event that the loan cannot be filled. Classified impact statements can be submitted via secure phone communications to USSOCOM SOAL-J-4, Comm: (813) 826-4275, DSN: 299-4275.

B-12. Component commands and TSOCs will review JOS loan requests to ensure compliance with training programs and operational requirements. Following their review, loan requests are then submitted to HQ, USSOCOM (ATTN: SOAL-J-4):

Mailing Address:
U.S. Special Operations Command
7701 Tampa Point Boulevard
ATTN: SOAL-J-4
MacDill AFB, FL 33621-5323

Phone/Fax Numbers:
DSN: 299-4275/9146
Comm: (813) 826-4275/9146
DSN Fax: 299-4741
Comm Fax: (813) 826-4741
Secure telephone unit (STU) III DSN Fax: 299-3780
STU III Comm Fax: (813) 828-3780

B-13. After the JOS loan request is received at USSOCOM HQ, the Center for Special Operations (SCSO-J-3) will validate the requirement and then pass it to the Directorate of Logistics (SOAL-J-4) for action, as appropriate. In cases where there are insufficient assets to meet requirements, the Directorate of Operations will deconflict, coordinate, and prioritize competing requirements.

B-14. Once the Directorate of Logistics receives the validated JOS loan request, they in turn will pass it on to SOFSA. SOFSA Special Programs Office then prepares the equipment for shipment IAW required delivery dates. SOFSA also contacts the requesting unit to advise them of the shipment and to ensure that the shipment is received. SOFSA acts as USSOCOM's agent and is the focal point for JOS loans once a JOS loan request is passed to them for action.

B-15. SOFSA will include a packing list, a memorandum of agreement (MOA), and a property accountability report with each JOS loan shipment as follows:
- The packing list will indicate stock number, nomenclature, and quantity shipped information.
- The MOA specifies the conditions under which the loan is made and directions for loan closure.
- The property accountability report (hand receipt) lists all items on the loan along with their serial numbers.

B-16. Units will return signed copies of the MOA and Property Accountability Report to SOFSA. These documents will be used by SOFSA to track accountability of JOS equipment on loan. Accountability is an important part of the JOS program and a precondition for units using the equipment. A report listing all JOS loans for which SOFSA has not received signed documentation is sent weekly to each component command.

Note: The JOS Loan Request Worksheet and JOS Communications Security (COMSEC) Loan Request Worksheet can be obtained on the USASOC home page under the G-4 tab.

Appendix B

JOINT OPERATIONAL STOCKS EQUIPMENT READINESS

B-17. JOS equipment is regularly serviced and maintained in a ready-to-ship status to meet time-sensitive operational contingencies and other emergency situations. Normally, units should allow as much processing time as possible and integrate JOS loan requirements into their planning. When submitting loan requests, units should also consider the amount of time it will take a request to be processed and approved by their own chain of command. However, SOFSA will maintain a capability to ship JOS equipment in 4 hours or less. This means SOFSA will prepare items for shipment, arrange for shipping, and have equipment turned over to the shipper in less than 4 hours during normal duty hours. During nonduty hours, SOFSA will recall necessary personnel to initiate this action. The exercise of the 4-hour response capability is restricted to national emergencies, contingency operations, and other urgent situations when authorized by HQ, USSOCOM.

SHIPMENT OF JOINT OPERATIONAL STOCKS EQUIPMENT

B-18. While USSOCOM budgets for the cost of shipping JOS equipment to the using unit, units are responsible for transportation costs associated with returning JOS equipment to SOFSA. If there is a significant operational impact, units must coordinate with USSOCOM Acquisition and Logistics for resolution. When large shipments of JOS equipment are intended for deployment overseas, the equipment will normally be shipped to the CONUS base where requesting units will assume responsibility for the equipment and arrange further shipping. These shipments will normally be managed as part of the unit's deployment and charged to the associated charge code (TAC Code) for the exercise or operation.

PROPERTY ACCOUNTABILITY

B-19. Commands/organizations using JOS-loaned equipment must maintain accountability over equipment on loan from time of receipt until the materiel is returned. Units are directed to maintain control of JOS loans through unit PBOs or equipment custodians as applicable.

B-20. JOS users will receive and maintain JOS equipment IAW the hand receipt and MOA that accompany each JOS loan. Users are further required to assign an accountable person to track JOS loans and to maintain property accountability records of equipment from the time of receipt until the property is returned. This person is normally the PBO or equipment custodian.

B-21. In the event JOS equipment cannot be returned IAW the JOS loan timelines, users are required to generate a loan extension request and to update their hand receipts if the extension is approved.

B-22. All users of JOS equipment will notify SOAL immediately if JOS equipment is lost or destroyed.

B-23. Units losing JOS equipment will initiate Reports of Survey, property adjustment documents, and investigations, as required. For loss or damage of JOS equipment, the following applies:
- A Report of Survey (DD Form 200 [Financial Liability Investigation of Property Loss]) will be initiated by the using command/organization to investigate and document the circumstances surrounding the loss, damage, or destruction of JOS equipment.
- The Report of Survey will be conducted IAW Service regulations. Upon completion by the Service reviewing authority, the survey will be forwarded to USSOCOM SOAL-J-4. USSOCOM SOAL-J-4 will provide SOFSA with guidance needed to adjust accountable records.
- For loss of sensitive items such as weapons and COMSEC, users will initiate reports and investigations as required by Service regulations.

RETURN SHIPMENT OF JOINT OPERATIONAL STOCKS EQUIPMENT

B-24. While the JOS program funds outbound transportation of JOS equipment, the user is responsible for the cost of returning JOS equipment to SOFSA. Return shipping information is as follows:

Shipping Address:
Special Operations Forces Support Activity
5749 Briar Hill Road
Building 221-SPO
ATTN: Joint Operational Stocks
Lexington, KY 40516

Phone/Fax Numbers:
DSN: 745-3805
Comm: (859) 293-3805
DSN Fax: 745-3899
Comm Fax: (859) 293-3899

Identification Numbers:
DODAAC: H92227
Unit Identification Code for SOFSA: DJ7511

B-25. The return of JOS COMSEC equipment must first be coordinated with the following personnel prior to shipment (normal duty hours are 0700 hours to 1530 hours Eastern Standard Time):

Shipping Address:
Special Operations Forces Support Activity
5749 Briar Hill Road
Bldg. 221-SPO
ATTN: COMSEC Custodian
Account Number: 871720
Lexington, KY 40516

Phone/Fax Numbers:
COMSEC Custodian: Ext 4214
DSN: 745-3945
Comm: (859) 293-3945
DSN Fax: 745-4169
Comm Fax: (859) 293-4169

B-26. JOS equipment returned to SOFSA will be complete, properly maintained, and serviceable. Unserviceable equipment will be returned with an explanation to allow for expeditious repair and restockage of the equipment.

B-27. For selected items of equipment, such as tents, requesting units are required to clean, inventory, and repack equipment as a condition of the loan to minimize high reconstitution costs.

TURN-IN GUIDANCE FOR UNSERVICEABLE JOINT OPERATIONAL STOCKS EQUIPMENT

B-28. This guidance applies to deployed SOF units and provides procedures for processing turn-ins of USSOCOM-procured (MFP-11) equipment to forward to SSA and Defense Reutilization and Marketing Services.

Appendix B

B-29. USSOCOM units are authorized to turn-in unserviceable JOS equipment to SSA and Defense Reutilization and Marketing Services, except for weapons, sensitive items, and COMSEC items. Units must obtain turn-in documentation, most commonly a DD Form 1348-1A (Issue Release/Receipt Document), identifying the item turned in, to include any serial numbers. The JOS loan number associated with the property turn-in should be included. A copy of this document must be forwarded to USSOCOM SOAL-J-4 to have the item cleared from the JOS loan.

B-30. This guidance does not apply to COMSEC items, weapons, and sensitive items such as special operations laser markers. JOS equipment falling under this limitation will be returned to the SOFSA, ATTN: JOS. Units should contact USSOCOM SOAL-J-4 for guidance before turning in unserviceable JOS equipment.

FAILURE TO RETURN JOINT OPERATIONAL STOCKS EQUIPMENT/FAILURE TO UPDATE HAND RECEIPTS

B-31. The JOS Loan Status Report and the JOS Property Accountability Report (Hand Receipts) are sent out weekly to component commands for action. It reflects the status of all JOS loans assigned to that command and the status of hand receipts requiring update. Component commands are requested to assist in getting overdue loans turned in. Units failing to return equipment in a timely or complete manner may be restricted from access to JOS except for operational requirements. Generally, unit accounts will be temporarily frozen when they have overdue loans 30 days beyond the due date or any unsigned hand receipts beyond 14 days from receipt of equipment or loan extensions. Accounts will remain frozen until the unit corrects its deficiencies.

Appendix C
Classes and Subclasses of Supply

This appendix provides a breakdown and description of classes and subclasses of supplies (Figure C-1, pages C-1 through C-4).

DESCRIPTION	CLASS
Subsistence: Potable water and food to sustain life.	I
Clothing, tools, supplies: Individual equipment, tentage, organizational tool kits, hand tools, maps, and administrative and housekeeping supplies and equipment. Includes items of equipment (other than principal items) prescribed in authorization and allowance tables and items of supply (not including repair parts).	II
POL: Petroleum fuels, lubricants, hydraulic and insulating oils, preservatives, liquid and compressed gases, bulk chemical products, coolants, deicing and antifreeze compounds (together with components and additives of such products), and coal.	III
Construction materiel: Installed equipment and all fortification and barrier materials.	IV
Ammunition: All types (including chemical, radiological, and special weapons), bombs, explosives, land mines, fuzes, detonators, pyrotechnics, missiles, rockets, propellants, and other associated items.	V
Personal demand items: Nonmilitary sales items (sundry packages).	VI
Major end items: Final combination of end items that are ready for their intended use—for example, tanks, launchers, mobile machine shops, and vehicles.	VII
Medical materiel: Medical supplies, repair parts, and equipment.	VIII
Repair parts (less medical-peculiar repair parts). All repair parts and components, including kits, assemblies and subassemblies, and repairable and nonrepairable items required for maintenance support of all equipment.	IX
Materiel to support nonmilitary programs: Agricultural and economic development materials not included in Classes I through IX.	X
DESCRIPTION	SUBCLASS
Air (aviation, aircraft, airdrop equipment):	A
• Class I—Food packet, in-flight, individual.	
• Class II—Items of supply and equipment in support of aviation and aircraft.	
• Class III—Petroleum and chemical products used in support of aircraft.	
• Class V—Munitions delivered by aircraft or aircraft weapons systems.	
• Class VII—Major end items of aviation equipment.	
• Class IX—Aircraft repair parts.	

Figure C-1. Description of classes and subclasses of supplies

Appendix C

DESCRIPTION	SUBCLASS
Troop support materiel: Water purification sets; shower, bath, laundry, dry cleaning, and bakery equipment; sets, kits, and outfits (including tool and equipment sets and shop and equipment sets) for performing organization, direct support, general support, and depot-level maintenance operations; sensors and interior intrusion devices; and topographic equipment and related topographic products as outlined in AR 115-11, *Geospatial Information and Services*.	B
Operational rations: Standard B rations, which are used for group feeding in areas where kitchen facilities, except refrigeration, are available and ration-supplement sundry packs are issued with the standard B rations until normal post exchange facilities are provided.	C
Commercial vehicles: Wheeled vehicles authorized for use in administrative or tactical operations.	D
General supply items: Administrative expendable supplies, such as typewriter ribbons, paper, cleaning materials, and other supplies normally referred to as office supplies. Also includes publications distributed through adjutant general channels.	E
Clothing and textiles: Individual and organizational items of clothing and equipment authorized in allowance tables and tentage or tarpaulins authorized.	F
Communications-electronics: Signal items, such as radio, telephone, satellite, avionics, marine communications, and navigational equipment; tactical and nontactical automated data-processing equipment; radar; photographic, audiovisual, and television equipment; and electronic sensors.	G
Test, measurement, and diagnostic equipment: Equipment used to determine the operating efficiency of—or to diagnose incipient problems in—systems, components, assemblies, and subassemblies of material used by the Army.	H
Tactical vehicles: Trucks, truck tractors, trailers, semitrailers, and personnel carriers.	K
Missiles: Classes II, VII, and IX, including guided missile and rocket systems, such as Patriot and Avenger. Class V includes guided-missile ammunition items.	L
Weapons: Small arms, artillery, fire-control systems, rocket launchers, machine guns, air defense weapons, and aircraft weapons subsystems.	M
Special weapons: • Class V—Nuclear and thermonuclear munitions. • Class VII—Weapons systems that deliver nuclear munitions. • Class IX—Repair parts for Class VIII-N.	N
Combat vehicles: Main battle tanks, recovery vehicles, self-propelled artillery, armored cars, and tracked and half-tracked vehicles.	O
U.S. Army Intelligence and Security Command materiel: Materiel for which the United States Army Intelligence and Security Command (INSCOM) has responsibility and which is normally authorized in classified authorization tables. Although INSCOM items are electronic, they are identified separately because they do not follow the supply and maintenance channels as subclass G.	P
Marine equipment: Marine items of supply and equipment, such as amphibious vehicles, landing craft, barges, tugs, floating cranes, and dredges.	Q

Figure C-1. Description of classes and subclasses of supplies (continued)

Classes and Subclasses of Supply

DESCRIPTION	SUBCLASS
Refrigerated subsistence: Two categories of refrigeration—that which is required to be maintained at 0 degrees Fahrenheit to keep frozen meals and foods for extended periods and that which is to be maintained at approximately 40 degrees Fahrenheit to keep perishables in A rations (such as fruits, vegetables, and eggs) for shorter periods.	R
Nonrefrigerated subsistence: Items in standard B rations and nonperishable items in A rations.	S
Industrial supplies: Common supplies and repair parts, such as shop stocks, hardware, and fabrication-type items generally having multiple uses. The Defense Industrial Supply Center generally manages such items.	T
Communications security materiel: This subclass is identified separately from subclass G because of specialized supply and maintenance functions performed through a dedicated COMSEC logistics system.	U
Ground: • Class I—Water, when delivered as a supply item. • Class II—Petroleum and chemical products and solid fuels used in support of ground and marine equipment. • Class V—Conventional munitions, such as chemical, smoke, illuminating, incendiary, riot-control, and improved conventional munitions. • Classes II, VII, and IX—Construction and road-building and materials-handling equipment.	W
In class: Indicates no subclass assigned.	X
Railway equipment: Rail items of supply and equipment, such as locomotives, railcars, rails, and rail-joining and rail-shifting equipment.	Y
Chemical: • Class II—Battle dress overgarments, M256 chemical detector kits. • Class VII—Protective masks and smoke generators. • Class IX—Protective mask filters, protective mask carriers, or individual decontamination kits.	Z
The Following Subclasses Also Apply for Class III	
Air, bulk fuels: Jet fuels and aviation gasolines, normally transported by pipeline, rail tank car, tank truck, barge, coastal or oceangoing tankers, and stored in a tank or container having a fill capacity greater than 500 gallons.	1
Air, packaged bulk fuels: Aircraft-unique petroleum and chemical products consisting generally of lubricating oils, greases, and specialty items normally packaged by the manufacturer and procured, stored, transported, and issued in containers or packages of 55-gallon capacity or less.	2
Ground, bulk fuels: Motor gasoline, diesel, kerosene, and heating oils normally transported by pipeline, rail tank car, tank truck, barge, coastal or oceangoing tankers, and stored in tank or container having a fill capacity greater than 500 gallons.	4

Figure C-1. Description of classes and subclasses of supplies (continued)

Appendix C

DESCRIPTION	SUBCLASS
Ground, packaged bulk fuels: Ground bulk fuels, which (because of operational necessity) are generally packaged and supplied in containers of 5- to 55-gallon capacity, except fuels in military collapsible containers of 500 gallons or less that are also considered packaged fuels.	5
Ground, packaged petroleum: Petroleum and chemical products, generally lubricating oils, greases, and specialty items normally packaged by the manufacturer and procured, stored, transported, and issued in containers of 55-gallon capacity or less.	6
Ground, solid fuels: Coal, coal heating tables, or bar.	7
The Following Subclasses Also Apply for Class VIII	
Controlled substances.	1
Tax-free alcohol.	2
Precious metals.	3
Nonexpendable medical items, not restricted.	4
Expendable medical items, not restricted.	5
Commander-designated controlled items.	6–9
U.S. Army Medical Materiel Agency-controlled sensitive items.	0

Figure C-1. Description of classes and subclasses of supplies (continued)

Appendix D
Site Survey Checklist

This appendix provides a sample site survey checklist (Figure D-1, pages D-1 through D-5). The checklist contains essential considerations in selecting a site.

S-1 CONSIDERATIONS
Mailing address, E-mail address, existing mail facilities. (Is DA Form 285 [Technical Report of U.S. Army Ground Accident] required?)
Communications for rapid contact of individuals in case of an emergency. (Red Cross, Embassy, telephone number, E-mail.)
Transportation available for emergency leaves. (Time and distance from nearest airport.)
Availability of post exchange sources and vans.
Laundry facilities.
Availability of religious services; existing religious facilities.
Public information coordination. (Embassy, Public Affairs Office.)
Availability of special services and recreation items.
MEDICAL EVACUATION CONSIDERATIONS
Primary methods of evacuation.
• Unit designation.
• Location.
• Phone numbers (DSN and commercial prefixes).
• Radio frequency and call sign.
• Response time to area.
• Special capabilities (stokes, winch, jungle penetrator).
• Limitations (night flying, altitude).
• Type vehicle, aircraft (for patient load, flying time).
• Format for requests.
Alternate methods of evacuation.
• Unit designation.
• Location.
• Phone numbers (DSN and commercial prefixes).
• Radio frequency and call sign.
• Response time to area.
• Special capabilities (stokes, winch, jungle penetrator).
• Limitations (night flying, altitude).
• Type of vehicle or aircraft (for patient load, flying time).
• Format for requests.

Figure D-1. Sample site survey checklist

Appendix D

MEDICAL EVACUATION CONSIDERATIONS (CONTINUED)

Other agencies to contact for evacuation (range control, hospital, rescue squad).
- Unit.
- Contact procedures.

On-site medical treatment facility.

Unit providing medical support (drop zone operations, ambulance).
- Service period.
- Point of contact (POC) and contact procedures.

SERVICING HOSPITAL FACILITIES CONSIDERATIONS

Nearest servicing hospital facilities (military and civilian).
- Name.
- Location.
- Size and capabilities (limits to care available).
- Contact procedures, including POC and phone number.

Other facilities available (if applicable).
- Dental.
- Preventive medicine.
- Veterinary.
- Class VIII supply.

S-2 AND S-3 CONSIDERATIONS

JSOTF or exercise director's HQ.
- Isolation area.
  - Billets.
  - Mess facilities and ration cycle.
  - Briefing area.
  - Recreation area.
  - Physical security.
- Operations center.
  - Office space.
  - Briefing area.
- Administrative center.
  - Warehouse and storage procedures.
  - Office space.
- General facilities.
  - Telephones.
  - Transportation.
  - Billets.
  - Mess facilities and water source.
  - Electricity.
  - Latrines.
  - Staging area.

Figure D-1. Sample site survey checklist (continued)

S-2 AND S-3 CONSIDERATIONS (CONTINUED)

Maneuver area.
- Military (leased or owned); how large?
- Civilian property; how large?
- State and federal forests; how large?
- Contracts for obtaining maneuver rights.
- Resources in maneuver area.
  - Map and aerial photo coverage.
  - Landing fields available; capabilities, restrictions, and procedures for use.
  - Open areas of drop zones and landing zones; size, number, restrictions on use. Is a drop zone survey (AF IMT Form 3823 (Drop Zone Survey)) or a landing zone survey (AF IMT Form 3822 (Landing Zone Survey)) available?
  - Type of terrain; will it support U.S. operations?
  - Weather; prevailing wind conditions for airborne operations.
  - Water conditions for underwater operations; tidal data.
  - Survival capability; water sources.
  - Restrictions in maneuver area.
- Environmental assessment.
  - Use of demolitions.
  - Use of timber for constructing shelters.
  - Digging of dumps, latrines.
  - Hunting and fishing restrictions.
  - Population centers and civilians living in maneuver areas; off-limits areas.
  - Approval to target civilian facilities. Prior coordination required.
  - Stability operations and guerrilla troop support.
- What forces are available?
  - Strength.
  - Status of training.
  - Dates available.
- What type of mission command relationship is to be established?
- What types of training are support elements interested in?
- Can support elements provide their own—
  - Transportation?
  - Rations?
  - Ammunition and air items?
  - Riggers?
  - Necessary items of individual equipment?
- Name, organization, and telephone number of contact at support element.

S-4 CONSIDERATIONS

Class I – Rations. (POC telephone number.)
- Type.
- Quantity.
- Location of pickup point.

Figure D-1. Sample site survey checklist (continued)

Appendix D

S-4 CONSIDERATIONS (CONTINUED)
• Contact, supporting unit. • Are signature cards required? • Are funds required? How much? Class II and IV – Repair parts, expendable and nonexpendable supplies. (POC telephone number.) • Contact for supporting unit. • Requisitioning procedures. • Type clothing required. Can support unit provide necessary seasonal clothing? If not, has request been sent to G-4? • Special non-table-of-organization equipment required. • Amount of funds required. • Target construction materials and estimated cost. Class III – POL. (POC telephone number.) • Contact for supporting unit. • Signature cards. Required? • Funds required. Amount? • Safety requirements. • Reimbursement method. Self-service supply center. (POC telephone number.) • Location. • Signature cards. • Funds. • Type of SSSC requirements (Class I supplies). • List of stockage items. Rigger. (POC telephone number.) • Are rigger facilities available? • Are repair items available? • What is the method of reimbursement for rigger end items? • Are parachute shakeout or drying facilities available? • Are adequate parachute-storage facilities to support redeployment available? Billeting. (POC telephone number.) • Are sufficient billets available? • Where are the billets located? • Are bedding and bunks available? • Are tents and cots available? • Are special quarters available? • Is precoordination required for heating, power, or running water in billets? Transportation. (POC telephone number.) • What are the types and quantities of vehicles required or obtained? • What is the supporting unit designation (for vehicle repairs)?

Figure D-1. Sample site survey checklist (continued)

Site Survey Checklist

S-4 CONSIDERATIONS (CONTINUED)

- Are Fort Bragg military drivers' licenses valid?
- Are DA Forms 348 (Equipment Operator's Qualification Record [Except Aircraft]) required?
- Can vehicles be kept out of motor pools for extended periods?
- Are forklifts, cranes, flatbeds, and buses available?
- Is A/DACG support available?
- Who is the POC for A/DACG support?
- Can the local airfield provide 463L pallets, dunnage, and nets?

Air support. (POC telephone number.)
- Type of aircraft.
- Dates and times available.
- Mission and local restrictions. Airmobile transportation and MEDEVAC.
- Ramp space for aircraft.
- Maintenance facilities and availability of repair parts.

Funds. (POC telephone number.)
- Funds required (total amount and breakdown by specific areas of items).
- Orders for Class "A" agents.

COMMUNICATIONS-ELECTRONICS CONSIDERATIONS

Power sources.
- Direct current power.
- Converter.
- Battery supply.
- Alternating current power.
- Voltage.
- Wattage.
- Phase.
- Cycle of power.

Sites available.
- Antenna field.
- Distance to airfield.
- Telephone facilities.

Equipment support.
- Quantities and types of radios required and available.
- Wire and field telephones required and available.
- Maintenance available, with and without fund cite.

Availability of frequencies.

Figure D-1. Sample site survey checklist (continued)

This page intentionally left blank.

Appendix E
Statement of Requirements Format

The SOR identifies and consolidates in priority all unit requirements that exceed organic capabilities. As shown in the outline format below (Figure E-1, pages E-1 through E-11), a complete SOR addresses in detail all aspects of logistics issues.

E-1. The SOR in direct support of designated operations is submitted using the subordinate unit's SOR process and procedures. Upon receipt of a mission or task, the unit conducts a thorough mission analysis using the military decisionmaking process; the SOR format below can also be used during the analysis process. As a result of this analysis, the unit determines materiel and nonmateriel requirements needed to accomplish assigned operational tasks. The unit then prepares the SOR, which identifies those requirements the unit cannot satisfy with its organic assets or capabilities (spreadsheets with tabs are available to document each requirement by service/class of supply). The unit then staffs the SOR and submits it through designated command channels. The SOR should also be staffed with the SB(SO)(A) ALE/ASPO and the supporting ASCC theater sustainment structure if the requirement can be satisfied in the theater. The key to the SOR process is to identify the need early on in the planning process to ensure the timely resourcing of the requirement. There are two types of SORs—the deployment statement of requirement (DSOR) and the pre-mission training statement of requirement (PMT SOR). Once deployed, there are other means to acquire what is needed for mission accomplishment (C-MNS, ESD/ONS, HNS/ACSA, and so on).

E-2. Funding (resourcing) for the SOR may come from programmed dollars and/or be submitted as an unfinanced requirement. In general, the command's fiscal year general and specific guidance will dictate how SORs will be funded. The preferred staffing approval guidance for supplemental (Global Operations Against Terrorist Networks) requirements is for the ARSOF commanding general—through the review and approval process of the cost estimate—to delegate the approval and validation of the SOR to the subordinate commands.

> *Note*: A SOR is not an authorization document. Commanders are authorized to obtain items of equipment (either through requisition or purchase) specifically listed in valid and applicable authorization documents, such as the MTOE, Table of Distribution and Allowance, or common table of allowance (CTA), provided adequate funding is available. A complete listing of Army authorization documents is contained in AR 71-32, *Force Development and Documentation – Consolidated Policies*, paragraph 6-3.

CLASSIFICATION

1. REFERENCES.
2. GENERAL.
    a. Supported unit.
    b. Time of support.
    c. Location of supported unit at time of support.
    d. Unit POCs.
    e. Database.
    f. Number of supported personnel.
    g. Force activity designator.

CLASSIFICATION

Figure E-1. Statement of requirement format

Appendix E

CLASSIFICATION

3. CONCEPT OF OPERATIONS.
   a. Mission. State the general mission of the unit, command, or operation.
   b. Desired Results. Provide a concise statement of the desired results of the requested support.
4. ASSUMPTIONS. Give the conditions that are likely to exist or that must exist for the support to be required. Relate the assumptions to specific requirements, as appropriate.
5. CONSTRAINTS. Define the situation that, if experienced, will degrade operations. Give conditions to specific requirements, as appropriate.
6. COMMAND, CONTROL, AND COORDINATION. Describe functional mission command of the unit. Attach an organizational diagram, if necessary, and describe the location of liaison with the HN.
7. SUPPLIES.
   a. Class I.
      (1) Requirements of Dining Facility. Identify personnel requirements.
      (2) Type of Dining Facility. Determine if the dining facilities should be one of the following:
         (a) U.S. Government.
         (b) U.S. civilian-contracted.
         (c) HN civilian-contracted.
         (d) HN military.
      (3) Augmentation. If dining facilities are U.S. Government facilities, identify the requirements for augmenting personnel.
      (4) Food Storage Facilities. Determine which of the following food storage facilities are required to contain a 30-day supply of rations.
         (a) Dry space in cubic feet.
         (b) Chill space in cubic feet.
         (c) Freezer space in cubic feet.
      (5) Insulated Food Containers. Determine the requirement for insulated food containers. List the number of containers and required meals.
      (6) Sack Lunches. Determine the requirements for sack lunches.
      (7) Meal Payment. Determine how individuals will pay for their meals.
         (a) Cash collection.
         (b) Payroll deduction.
         (c) Meal cards.
      (8) Dining Facility Hours. Determine the requirement for a 24-hour facility.
      (9) Equipment Augmentation. Determine the requirement for equipment augmentation. List the equipment by nomenclature, National Stock Number (NSN), and quantity.
      (10) Combat Rations. Estimate the number of combat rations for 30-day sustainment packages for aviation personnel.
         (a) MREs.
         (b) Long-range reconnaissance patrol rations.
         (c) Other (specify).
      (11) Pre-positioned Rations. List the number of days required for pre-positioned rations.
      (12) Percentage of Pre-positioned Rations. Identify the required percentage of the following pre-positioned rations:
         (a) MREs.

CLASSIFICATION

Figure E-1. Statement of requirement format (continued)

CLASSIFICATION

  (b) Long-range reconnaissance patrol rations.
  (c) Other (specify).
 (13) Local Purchasing, Cash. Determine the need for small units to have cash to purchase rations on the economy.
b. Class II.
 (1) Self-Service. List the essential SSSC items required for a 30-day sustainment.
 (2) CBRN Equipment. List the requirement for CBRN consumables and nonconsumables for two complete issues of CBRN equipment following a CBRN attack.
 (3) Sustainment. List other Class II items required for sustainment, such as CTA 50-900 items.
 (4) Reproduction Equipment. Determine the required reproduction equipment. List the equipment and the number of copies needed for 30-day sustainment.
 (5) Special Equipment. List any special Class II equipment required beyond the equipment already authorized and on hand. List the equipment by nomenclature, NSN, and quantity.
 (6) Clothing Sales. Determine the requirement for a clothing sales facility.
c. Class III.
 (1) POL. Determine POL, including base support functions, for a 30-day sustainment. List item by type and quantity.
  (a) Motor gasoline (regular or super).
  (b) Diesel fuel.
  (c) Aviation gasoline.
  (d) Oil (bulk).
  (e) Grease.
  (f) Coolants.
  (g) Packaged POL or other lubricants.
  (h) Commercial grades of aviation fuel.
  (i) Availability of POL laboratory for fuel testing.
 (2) Tankers and Dispensers. Identify the requirement for tankers or dispensers in addition to organic capabilities. List item by type, capacity, and quantity.
 (3) Planning Factors. Determine if the planning factors used to identify POL requirements were factors other than those in the Combined Arms Support Command database or operational log planner. If so, specify.
d. Class IV. Determine the requirements for building or barrier materials for the following items, listing them by type and quantity—for example, plywood, lumber, long and short pickets, sandbags, and barbed wire.
 (1) Administrative and command post.
 (2) Tactical and defensive use.
 (3) Rigging and shoring.
 (4) Concertina and barbed wire.
 (5) Stakes and pickets.

CLASSIFICATION

Figure E-1. Statement of requirement format (continued)

Appendix E

CLASSIFICATION

e. Class V.
  (1) Additional Class V Requirements. Determine Class V requirements beyond those in the unit basic load. List by DOD identification code, nomenclature, and quantity. Identify nonstandard Class V requirements separately.
  (2) Planning Factors. Determine the planning factor used to forecast Class V consumption rates.
f. Class VI. Determine the number of personal demand items based on the number of unit personnel and the individual consumption rate. List items by type and quantity.
g. Class VII.
  (1) Additional Equipment. Determine the requirement for additional items of equipment, such as trucks and generators. List the items by nomenclature, NSN, and quantity.
  (2) Maintenance Augmentation. Determine the requirement for augmenting maintenance personnel to support the equipment listed in paragraph 7g(1). List the personnel by grade, MOS, and quantity.
h. Class VIII.
  (1) Determine the requirement for Class VIII supplies by nomenclature, NSN, quantities, and special requirements associated with a particular item, such as refrigeration.
  (2) Determine the schedule of resupplies required.
  (3) Determine whether resupply will be prepackaged standard line items. Project when line item ordering will be established and, if feasible, how often, how long, and through what channels.
  (4) Determine the need for Class VIII supplies peculiar to the AO and whether the supplies are readily available or must be specifically acquired, such as refrigeration, security, and shelf life—for example, antivenins.
  (5) Determine the availability and reliability of HN Class VIII for emergency purposes.
  (6) Determine the need for blood and blood products and associated equipment.
  (7) Determine coordinating agencies for chemical support.
i. Class IX.
  (1) Mandatory Parts List. Determine if a mandatory parts list exists to support the equipment.
  (2) Prescribed Load List. Determine if prescribed load list includes repair parts to support—
    (a) Weapons.
    (b) Communications equipment.
    (c) Vehicles.
    (d) Support equipment, such as generators.
    (e) CBRN equipment.
  (3) Other Equipment. Determine if the unit has nonstandard or commercial equipment. List by type, model number, manufacturer, and density.
  (4) Repair Parts Support. Identify how repair parts support is obtained for commercial and nonstandard equipment.
  (5) Maintenance Support. Determine maintenance support requirements.
j. Class X. Determine Class X requirements. List by type and quantity.
k. Other.

CLASSIFICATION

Figure E-1. Statement of requirement format (continued)

CLASSIFICATION

(1) Emergency Resupply. Identify the requirement for emergency prepackaged resupply. Specify by nomenclature, NSN, and quantity. Attach as separate enclosure for each type of package.
(2) Maps and Photographs. Identify the requirement for maps and aerial photographs.

8. SERVICES.
   a. Field Services. Determine field services support requirements, such as riggers and mortuary affairs.
   b. Engineering Services.
      (1) Equipment Power Rating. Determine power rating needed for the equipment.
      (2) Power Requirements. Determine power requirements beyond the organic generating capability.
      (3) Equipment Power Capability. Determine the following, if supplied with commercial power at the wartime site:
         (a) Equipment compatibility.
         (b) Requirement for plug adapters, including voltage and the number of adapters needed.
         (c) Requirement for transformers, including voltage and the number of transformers needed.
      (4) Water Requirements. Identify daily requirements for potable water and for washing engines and fuselages.
      (5) Pest Control Requirements. Determine the requirement for rodent- and insect-control assistance.
      (6) Heavy Equipment Requirements. Identify requirements for heavy engineer equipment, such as bulldozers. List the needed quantity.
   c. Other Services.
      (1) Linen Requirements. List by type and quantity.
      (2) Linen Exchange. Determine the frequency of linen exchange.
      (3) Laundry Services Requirements. List by pounds per week. If none, so state.
      (4) Commercial Cleaning Requirements. Determine the requirement for commercial laundry and dry cleaning.
      (5) Other Services Identification. Determine the need for other services.
9. MAINTENANCE.
   a. Personnel Requirements. Determine if enough personnel exist to conduct the necessary maintenance. If not, list the necessary augmentation by grade, MOS, and quantity.
      (1) Vehicle.
      (2) Support equipment.
      (3) Communication.
      (4) Weapons.
      (5) Aviation.
   b. Field and Sustainment Maintenance. Identify requirements for field and sustainment maintenance.
   c. Other Maintenance Equipment. List commercial and nonstandard equipment requiring maintenance.

CLASSIFICATION

Figure E-1. Statement of requirement format (continued)

Appendix E

CLASSIFICATION

10. TRANSPORTATION.
    a. Air Transportation.
        (1) Unit Load Plans. Enclose unit load plans.
        (2) Administrative Aircraft. Determine the requirement for administrative aircraft. Specify the type and number of hours per week.
        (3) 463L Pallets. Determine the requirement for 463L pallets at the wartime location. Specify the amount.
        (4) Equipment and Personnel Requirements. Determine the requirement for additional MHE and personnel at the JSOTF and SOTF airfield. Specify requirement for crane or for rough terrain container handler.
        (5) Passenger Facilities. Determine the requirement for passenger facilities. Specify the required type and size of the facilities, based on the duration of passenger use.
        (6) Cargo Storage Facilities. Determine the requirement for cargo storage facilities. Specify by the number of square feet required for the following:
            (a) Covered secure storage.
            (b) Outdoor secure storage.
        (7) Airfield Requirements. Determine the requirement for an airfield to handle the following:
            (a) C-130s.
            (b) C-17s.
            (c) C-5As.
            (d) Other (specify).
        (8) All-Weather Surface Airfield. Determine the requirement for an all-weather surface airfield.
        (9) Airfield Services. Determine the requirement for airfield services, including MHE support. List by type and quantity.
        (10) Airfield Operations. Determine the requirement for airfield operations provided by other sources.
        (11) Flight Line Facilities. Determine the requirement for other aircraft flight line facilities. Specify the types of aircraft.
    b. Water Transportation. Determine the requirement for water transportation. Specify the type and size of the maritime vehicle.
    c. Ground Transportation. Determine the requirement for supplemental military vehicles. Specify by type of vehicle and quantity.
        (1) Commercial-Type Military Vehicles.
            (a) Sedan.
            (b) Carryall.
            (c) Bus.
            (d) Ambulance.
            (e) Other (specify).
        (2) Tactical Vehicles.
            (a) Radio-equipped vehicles (state type of radio and quantity).
            (b) Non-radio-equipped vehicles.
            (c) Trucks and trailers.

CLASSIFICATION

Figure E-1. Statement of requirement format (continued)

CLASSIFICATION

    (d) Wreckers and cranes.
    (e) Aircraft-towing vehicles.
    (f) Ambulances.
    (g) Fire trucks.
    (h) Other special purpose vehicles, such as warehouse trucks.

11. FACILITIES.
   a. Maintenance Facilities (list in square feet).
     (1) Maintenance Area Requirements. Identify vehicle maintenance area requirements.
      (a) Number of bays.
      (b) Number of pits.
      (c) Aircraft parking (concrete or asphalt).
      (d) Wash racks.
      (e) Secure storage (tools; test, measurement, and diagnostic equipment [TMDE]).
      (f) Secure storage (repair parts).
     (2) Signal Maintenance Area Requirements. Identify signal maintenance area requirements.
      (a) Power.
      (b) Safety.
      (c) Secure storage (repair parts).
      (d) Secure storage (tools, TMDE).
     (3) Weapons Maintenance Area Requirements. Identify weapons maintenance area requirements.
     (4) Aviation Maintenance Area Requirements. Identify aviation maintenance area (covered) requirements.
      (a) Aircraft parking (concrete or asphalt).
      (b) Secure storage (repair parts).
      (c) Secure storage (tools, TMDE).
   b. Billeting Facilities.
     (1) Billet number and size requirements. List the number of billets and required square feet.
      (a) Officers.
      (b) Senior enlisted.
      (c) Enlisted.
      (d) Females.
     (2) Tentage. Determine if sufficient tentage is available within the unit to house personnel. If not, specify number and types of tents, climate-controlled systems, if necessary, and date-time group, when required.
     (3) Showers. Determine the required number of showers.
     (4) Latrines. Determine the number and location of latrines.
   c. Medical Facilities. Determine the requirement for physical facilities and optometry.
     (1) Hospital beds.
     (2) Treatment rooms.
     (3) Dental treatment rooms.

CLASSIFICATION

Figure E-1. Statement of requirement format (continued)

Appendix E

CLASSIFICATION

   (4) Laboratories.
   (5) X-ray rooms.
   (6) Pharmacies.
   (7) Other (specify).
  d. Other Facilities (list by function and square feet).
   (1) Operations center.
   (2) Logistics center.
   (3) Signal center.
   (4) Reception and palletizing facilities.
   (5) Dining facility.
   (6) Dispensary.
   (7) Isolation facility.
   (8) Parachute rigging and drying facility.
   (9) Ammunition storage.
   (10) Clubs.
   (11) Gym.
   (12) Antenna fields.
   (13) Ranges (list types of weapons requiring ranges).
   (14) Drop zones.
   (15) Secure facilities (for storing, receiving, and transmitting classified messages and documents).
   (16) Other (specify).
12. PERSONNEL SERVICES.
  a. Personnel.
   (1) Military Occupational Specialty. Identify critical MOSs (include additional skill identifiers and special qualifications identifiers).
   (2) Personnel Action Dissemination. Determine the routing of the following personnel actions and the classification of those actions:
    (a) Assignments.
    (b) Reassignments.
    (c) Efficiency reports.
    (d) Awards.
    (e) Promotions.
    (f) Reclassifications.
    (g) Other.
   (3) Casualty Reporting. Determine the procedures for casualty reporting.
   (4) Civilian Personnel Requirements. Identify civilian personnel requirements.
  b. Administrative Services.
   (1) Reproduction and Word Processing. Determine reproduction and word processing requirements.
   (2) Equipment Requirements. Determine the requirement for administrative equipment beyond the present equipment. Specify the type and quantity of systems.

CLASSIFICATION

Figure E-1. Statement of requirement format (continued)

Statement of Requirements Format

CLASSIFICATION

 (3) Blank Forms and Publications. Determine the requirement for pre-positioning of blank forms and publications.
 (4) Accident Reporting Procedures. Determine the requirement for accident reporting procedures (DA Form 285) and other related safety reports.
 (5) Postal. Identify postal requirements.
 c. Finance Support. Determine and identify the type of finance support requirements:
  (1) Type of required currency (procurement, disbursing, accounting, banking, and currency support).
  (2) Casual payments, check cashing, travel pay processing, and local currency conversion.
  (3) Commercial vendor services.
 d. Religious Support.
  (1) Religious Support Requirements. Determine the following religious support requirements:
   (a) Catholic.
   (b) Protestant.
   (c) Jewish.
   (d) Orthodox.
   (e) Muslim.
   (f) Other.
  (2) Vehicular Support. Determine additional equipment support, including transportation and communications and computer systems, to accomplish the religious support mission.
 e. Legal. Determine the requirement for staff judge advocate support in the following areas:
  (1) Administrative law.
  (2) Claims.
  (3) Defense.
  (4) Prosecution.
  (5) International law.
  (6) Operational law, including rules of engagement.
 f. Public Affairs. Determine the requirement for public affairs office support.
13. ARMY HEALTH SYSTEM SUPPORT.
 a. Hospitalization (Theater Army or area of responsibility [AOR]).
  (1) Determine the patient estimate for the number of required hospital beds.
   (a) Surgical.
   (b) Medical.
  (2) Determine the location and accessibility of supporting Roles 3 and 4 hospitals.
 b. Medical Treatment and Patient Evacuation.
  (1) Medical Treatment. Roles 3 and 4 hospitals provide medical treatment. ARSOF must state their specific patient-tracking requirements.
  (2) Organic Support.
   (a) Availability of assets.

CLASSIFICATION

Figure E-1. Statement of requirement format (continued)

Appendix E

CLASSIFICATION

      (b) Casualty evacuation. This requirement includes using SOAR and Army evacuation assets to extract casualties from hostile and denied territory. It should also include augmentation of SOAR aircraft with medical personnel (physicians, physician assistants, or ARSOF medics), as required.

   (3) Theater Army or AOR Support. Use traditional assets to support MEDEVAC only if those assets do not compromise the security of the operation.

      (a) Ground evacuation assets in sustainment, field, or area-support roles.

      (b) Air evacuation assets in sustainment, field, or area-support roles.

   (4) Theater of Operations Evacuation Policy. A requirement may exist that an exception to the theater evacuation policy is necessary to retain qualified ARSOF personnel within the theater.

  c. Area Medical Support. The ARSOF must identify the location of the medical support organizations and project medical treatment beyond organic capabilities.

  d. Dental Services. Determine dental support requirements and location of support organizations.

  e. Preventive Medicine Services. Determine preventive medicine support requirements beyond organic assets and the location of support organizations.

  f. Veterinary Services. Requirements for veterinary support must be identified and coordinated through the appropriate mission command element.

  g. Combat and Operational Stress Control (COSC). Control of stress is a command's responsibility. The ARSOF must plan for COSC support and identify COSC organizations.

  h. Medical Logistics.

   (1) Determine requirements for Class VIII supplies by nomenclature, NSN, quantities, and special requirements associated with a particular item, such as refrigeration.

   (2) Determine schedule of resupplies required.

   (3) Determine whether resupply will be prepackaged standard line items. Project when line-item ordering will be established and, if feasible, how often, how long, and through what channels.

   (4) Determine the need for Class VIII supplies peculiar to the AO and whether they are readily available or must be specifically acquired, such as refrigeration, security, and shelf life (for example, antivenins).

   (5) Determine availability and reliability of HN Class VIII for emergency purposes.

   (6) Determine need for blood and blood products and associated equipment, if required.

   (7) Determine coordinating agencies for chemical support.

  i. Medical Laboratory Support. Determine medical laboratory support requirements and support organizations.

14. SIGNAL.

  a. Terminal Equipment and Access. Determine requirements for the following:

   (1) Supplemental terminal equipment. Specify by type and quantity.

   (2) Access to HN commercial telephone system. Specify need, such as number of lines.

   (3) Access to NATO telegraph network.

   (4) Access to HN military teletype system.

   (5) Access to automatic secure voice communications.

   (6) Access to NATO secure voice network.

   (7) Access to Automatic Digital Network.

CLASSIFICATION

Figure E-1. Statement of requirement format (continued)

CLASSIFICATION

  (8) Identify data communications requirements for standard Army multi-command management information system (STAMMIS) and other data systems. Specify intertheater and intratheater requirements.
 b. Transmit and Receive Sites. Determine the number of transmit and receive sites to be set up and the amount of area necessary.
  (1) Access to NATO telegraph network.
  (2) Access to HN military teletype system.
  (3) Access to automatic secure voice communications.
  (4) Access to NATO secure voice network.
  (5) Access to Automatic Digital Network.
  (6) Identify data communications requirements for STAMMIS and other data systems. Specify intertheater and intratheater requirements.
 c. Signal Maintenance Support. Determine the requirements for supplemental signal maintenance support.
 d. Frequency Requirements. Determine the number of separate frequencies needed daily.

15. SECURITY.
 a. Military Police Functions. Determine the requirement for the following military police functions:
  (1) Access control.
  (2) Detention (prisoner of war and friendly).
  (3) Investigations.
  (4) Traffic control.
  (5) Physical security.
  (6) General law enforcement.
  (7) Convoy security.
  (8) Special weapons.
  (9) Other (specify).
 b. Counterintelligence. Determine the requirement for counterintelligence.
 c. Base Defense. Determine the requirement for base defense capabilities.

16. FUNDING AND FINANCIAL MANAGEMENT SUPPORT.
 a. Resource management.
 b. Cost-capturing requirements.
 c. Procurement.
 d. Disbursing.
 e. Accounting.
 f. Banking.
 g. Currency support.

CLASSIFICATION

Figure E-1. Statement of requirement format (continued)

This page intentionally left blank.

Appendix F
Health Threat and Medical Intelligence

History provides examples of battles that were lost because troops were immobilized by disease and nonbattle injuries. A critical element of the AHS assessment is a thorough appraisal of the health threat. The assessment includes the health threat to the deploying forces and to the residents in the AO. AHS planners use a systematic process, called the medical intelligence preparation of the battlefield (MIPB), to analyze various enemy, environmental, and health threats in a specific AO. This appendix discusses the health threat, the health threat assessment, and the MIPB. It also provides a sample format for a medical intelligence support appendix to the intelligence annex of an OPLAN.

HEALTH THREAT

F-1. Health threats that account for the vast majority of combat ineffectiveness fall into five broad categories:
- *Environmental Injuries and Conditions.* This category includes heat and cold injuries resulting from inadequate acclimation to the AO, inadequate clothing and equipment for the environmental conditions, dehydration, and exposure to cold and wetness. This category may also include occupational hazards, such as carbon monoxide, toxic industrial chemicals, and noise.
- *Endemic and Epidemic Diseases in the AO.* This category includes diseases of military significance, diarrhea diseases caused by drinking contaminated or impure water (not adequately treated), eating contaminated foods, and not practicing good individual and unit preventive medicine. These diseases may also be the result of disease transmission by arthropod vectors.
- *Diseases and Injuries Caused by Contact With Animals and Plants.* This category includes contact with wild animals, domesticated animals, reptiles, and poisonous or toxic plants.
- *Diseases and Injuries Caused by Physical or Mental Unfitness.* This category includes conditions that may occur from continuous operations, fatigue, inadequate diet, and mental stresses.
- *Diseases and Injuries Resulting From CBRN Exposure.* This category includes exposure to CBRN warfare agents and weapons.

HEALTH THREAT ASSESSMENT

F-2. U.S. Soldiers are at high risk in stability operations, as the incidence and exposure to infectious diseases and environmental hazards are great in man-made or natural disaster areas and in developing nations. The health threat is derived through established intelligence channels and from a variety of informational sources outside the military.

F-3. The ability to obtain, interpret, and use medical intelligence is critical to the success of the AHS mission. Military operations and man-made and natural disasters can cause a resurgence of diseases once thought to be at low epidemiological levels and may also result in environmental contamination. A combination of factors can result in the spread of communicable diseases in epidemic proportions and increased opportunity for exposure to CBRN hazards. These factors are—
- Disruption of sanitation services, such as garbage disposal or sewer systems.
- Contamination of food and water.

Appendix F

- Development of new breeding grounds for rodents and arthropods, such as in rubble or in stagnant pools of water.
- Disruption of industrial operations.
- Dispersion of biological or radiological waste by improper handling or terrorist activity.

MEDICAL INTELLIGENCE PREPARATION OF THE BATTLEFIELD TEMPLATE

F-4. MIPB is a systematic process that aids AHS planners in analyzing various enemy, environmental, and health threats in a specific AO. The MIPB process is the first step in the mission analysis phase of the contingency-planning process. The information derived from conducting a proper MIPB is the cornerstone to developing detailed, effective AHS estimates and plans. Some portions of the template are more or less applicable, depending on the assigned mission. The purpose of MIPB is to—

- Define the battlefield environment.
- Describe the battlefield effects on deployed forces and AHS operations.
- Conduct threat integration (enemy and medical) and information consolidation.

F-5. A broad interpretation may be needed when applying categories to stability operations. Figure F-1, pages F-2 through F-6, describes an MIPB template.

---

1. **Battlefield Environment.**
    a. Identify significant characteristics of the environment.
        (1) Geography. Describe the climate, weather, terrain, and altitude. Include information on possible weather and environmental threats, such as earthquakes, volcanoes, and monsoons.
        (2) Political and socioeconomic situation. Describe population demographics, such as ethnic groups, religious groups, age distribution, income groups, culture, and language. Also, describe living conditions of the general population, infant mortality rate, anticipated requirements for medical support of the local population, and the refugee or displaced person situation. Describe the role of clans, tribes, gangs, opposition groups, and paramilitary organizations and groups, as well as crime rates and the presence of organized crime.
        (3) Threat forces and capabilities. Describe enemy ideology, goals, objectives, and missions, as well as the enemy's attitude toward the Geneva Conventions. Describe threat characteristics (in broad terms) and enemy force structure and weapons systems. Also, describe the enemy's capability to generate friendly casualties and the types of wounds or injuries anticipated. Describe enemy medical doctrine and capabilities. Indicate whether U.S. forces are likely to treat significant numbers of enemy wounded. Describe the overall health status of the enemy, including significant endemic and epidemic diseases present and immunization status. Describe CBRN weapons and agents, delivery systems, doctrine for use, and ability to sustain operations in a CBRN environment. Describe medical logistics structure, including quality, quantity, availability, and types of medical equipment. Also, describe enemy psychological operations and UW capability.
        (4) Infrastructure. Describe the infrastructure, including transportation systems (land, sea, and air), communications systems (telephone, cellular, digital, mass media, and electronic means), and utilities (water, electricity, and sanitation).

Figure F-1. Medical intelligence preparation of the battlefield template

(5) Medical infrastructure. Include the location and availability of medical facilities. Indicate the quality and types of medical facilities, including names and contact information for practitioners and health administrators. Describe the capabilities of medical facilities (size, patient capacity, and types of specialties) and the education and training levels of health services professionals and ancillary support personnel. Indicate whether enemy forces will use or have access to the civilian medical system and whether the medical facilities are approved facilities for use by U.S., allied, or coalition forces. Indicate the quality and availability of medical supplies, pharmaceuticals, blood, and blood products. Describe evacuation capability, services, and availability, and include names and contact information. Identify location of helipads, railheads, airheads, seaports, medical waste incinerators, disposal areas, and availability of contract support.

(6) Health threat. Describe endemic and epidemic diseases, as well as environmental injuries and conditions. Identify diseases and injuries from wild animals, domesticated animals, reptiles, and poisonous or toxic plants. Also, identify diseases and injuries from physical or mental unfitness and from exposure to CBRN agents.

(7) Nongovernmental organizations operating in the AO. Include such organizations as the International Committee of the Red Cross or Doctors Without Borders.

b. Identify the limits of the command AO. The command AO is the geographic area where the commander is assigned the responsibility and authority to conduct military operations.

(1) Identify the geographic AOR. Include the macro view or the micro view, depending on the level of command and the size of the geographic area.

(2) Identify the total population at risk. Include all U.S., allied, coalition, or HN forces; local civilian population; refugees and displaced persons; employees and contractors of the U.S. Government; and nongovernmental organization personnel. Determine individuals or groups eligible for health care provided by U.S. Army HSS assets.

(3) Identify all supported U.S. units. Include sister Services and elements from U.S. Government agencies and contractors.

(4) Identify all supported allied, coalition, HN, or other multinational units or elements. Discuss unit troop strengths, locations, and missions. Identify organic medical resources and capabilities. Also, identify multinational medical assets approved for use by U.S. personnel. Identify unique medical support requirements, such as endemic diseases, in allied and coalition forces that are not present in the deployment AO. Also, identify the current level of health and dental fitness among the supported populations.

c. Establish limits of the area of interest (AOI). The AOI is a geographic area from which information is required to facilitate planning. The AOI usually falls outside the AO and may or may not be applicable to a particular operation. The AOI is of concern when portions of the overall HSS plan fall outside the AO.

(1) Organizations and elements outside the AO that provide HSS. For example, CONUS support base hospitals, AHS support (Defense Logistics Agency or U.S. Army Medical Materiel Agency), and global patient regulating support (Global Patient Movement Requirements Center).

(2) Location and time-distance factors for HSS resources to augment, reinforce, or reconstitute HSS units or personnel within the AO. Include information on units or elements in the CONUS support base or adjacent theaters.

(3) Coordination and synchronization with mission command assets outside the AO.

(4) Follow-on operations or operations being conducted simultaneously outside the AO.

d. Identify the level of detail required and the time available to conduct MIPB.

Figure F-1. Medical intelligence preparation of the battlefield template (continued)

Appendix F

> e. Evaluate existing information and intelligence of medical significance and identify intelligence gaps. Sources include the National Center for Medical Intelligence, Defense Intelligence Agency, country studies, intelligence officers, operations and training officers, and military intelligence units. Other sources include the Central Intelligence Agency, tourist maps and brochures, preventive medicine resources, World Health Organization, Pan American Health Organization, DOS, the Internet, and libraries.
> f. Identify and submit collection requirements to support intelligence staff sections, elements, and units.
> g. Collect required information to fill gaps.
>
> **Note:** Should HSS personnel gain information of potential medical intelligence value while in the performance of their duty, they are required to report it to their supporting intelligence element.
>
> 2. **Battlefield Effects.** The purpose of this phase of the MIPB process is to analyze and integrate various factors of the battlefield environment. Detailed analysis of these factors, to determine the military significant effects, results in medical intelligence upon which the commander can make informed decisions. The emphasis is on the effects on friendly forces, as well as friendly and enemy actions.
>   a. Geography.
>   (1) Climate and weather effects on operations. Include the effects of extreme heat, cold, and humidity. Also, include the effects of predominant weather patterns, heavy rain or snow, and phases of the moon. Describe climatic effects on medical supplies and equipment and the effects of enemy chemical and biological agents on the weather.
>   (2) Terrain analysis. Determine effects of terrain on friendly and enemy maneuver capability and ability to sustain health care. Identify effects on timely medical evacuation, natural lines of patient drift, and medical treatment facility site-selection factors.
>   (3) Altitude effects. Identify effects of high-altitude operations on force capability and rotary-wing evacuation assets. Identify standard medical treatment protocols.
>   b. Political and socioeconomic situation.
>   (1) Population demographics. Include the effect on the delivery of HSS to supported forces and on the HSS system, if required to support the local populace or nongovernmental organizations. Identify the political effects of providing or not providing care to the HN populace, nongovernmental organizations, refugees, and displaced persons. Identify the effects of cultural, religious, or language barriers.
>   (2) Condition of the general population (or supported population). Include an analysis of the health of the general population and its impact on deployed forces. Identify the infant mortality rate, which is an indicator of the overall health of the population and the state of advancement of the medical system.
>   (3) Effects of clans, tribes, gangs, opposition groups, or paramilitary organizations or groups and organized crime on the ability to provide HSS to deployed forces and other eligible beneficiaries.
>   (4) Additional requirements of refugees, displaced persons, and enemy prisoners of war on the HSS system. These requirements are particularly important in preventive medicine, as camps require sanitation, pest management, and potable water support. Other requirements include provision of sick-call services, outpatient treatment, hospitalization, evacuation, and medical logistics support (such as sorting, repackaging, inventorying, and disseminating donated medical supplies and equipment).
>   c. Threat forces capabilities and effects.

Figure F-1. Medical intelligence preparation of the battlefield template (continued)

(1) Effects of enemy ideology, goals, and missions. Analyze the enemy's will to fight, his military objectives, and his compliance with the Geneva Conventions. Also, analyze the types of enemy forces (paramilitary, conventional, or SO); philosophy concerning collateral damage, civilian casualties, and disruption of utilities (sewage, waste disposal, sanitation, water, electricity, and gas); and creation of refugees or displaced persons.

(2) Threat characteristics. Identify the effects of enemy doctrine on deployed forces, including medical personnel and units. Identify friendly units, elements, and organizations most likely to sustain heavy casualties.

(3) Enemy force structure and weapons systems. Analyze the accuracy and range of enemy weapons systems, the size and composition of the enemy force, and the types of friendly wounds generated by enemy weapons systems (such as piercing, concussion, blunt trauma, or burns).

(4) Enemy medical doctrine and capabilities. Analyze enemy medical doctrine and capabilities. Analyze priority and availability of medical care and medical evacuation. Identify the infrastructure and training to accomplish the medical mission. Determine the potential for the enemy to treat its own casualties or to leave them for the care of friendly forces.

(5) Effects of enemy weapons of mass destruction. Analyze enemy weapons of mass destruction capabilities, the effects of enemy CBRN use on friendly forces, and the likelihood of the enemy use of weapons of mass destruction. Determine whether the enemy can continue the mission in a CBRN environment and if enemy delivery systems are accurate, reliable, and effective.

(6) Effects of Psychological Operations and UW. Analyze the probable impact of psychological operations on friendly forces. Also, analyze UW capabilities and the probability of UW forces targeting friendly rear area and HSS assets and resources. Identify the effect of UW on the delivery of health care.

d. Infrastructure.

(1) Transportation systems. Identify the effect of available transportation systems on timely patient evacuation, medical logistics supply and resupply operations, and enemy casualty evacuation. Also, identify the likely avenues of approach, effect of the transportation system on mobility and military operations, effect of military operations on the transportation system, and impact of transportation networks on enemy and friendly courses of action.

(2) Communication systems architecture.

(3) Utilities (water, electricity, and sanitation). Analyze water quality (potability) and distribution system. Identify the reliability of electrical power generation and the effectiveness and efficiency of sanitation systems. Determine the effects of enemy and friendly military actions on the utilities infrastructure and the impact of a disruption of utilities on the health of the general population and deployed forces.

e. Medical infrastructure.

(1) Indigenous medical facilities.

(2) Local sources of medical supplies. Analyze quantity, quality, and availability of local medical supplies and equipment. Analyze the availability of blood and blood products. Also, determine the availability of supplies for use by the local populace, refugees, enemy prisoners of war, and displaced persons. Determine the availability of supplies approved for use by U.S. forces. Analyze local medical supply production facilities, the impact of military operations on the local medical supply infrastructure, and the availability and quality of medicinal gases.

Figure F-1. Medical intelligence preparation of the battlefield template (continued)

Appendix F

> (3) MEDEVAC services. Analyze local MEDEVAC services and capabilities, coordination and synchronization of local evacuation services and resources to redirect civilian patients, availability and quality of local medical treatment facilities, and impact of military operations on local evacuation services.
>
> (4) Effects of disease and other environmental threats. Identify disease and environmental threats that affect friendly forces and the delivery of HSS. Also, identify preventive measures required to counter the health threat and the impact of disease and environmental threats on enemy actions. Identify additional disease and environmental hazards created or aggravated by military operations.
>
> f. Analysis of services provided by nongovernmental organizations.
>
> 3. **Threat Integration and Information Consolidation.** The objective of threat integration is to determine the effects of essential elements of friendly information (EEFI) on the health of the command, the employment of HSS resources, and enemy and friendly courses of action. Overlays, spreadsheets, matrices, and databases are useful formats for managing information and medical intelligence.
>
>   a. Threat integration can be broken down into two major categories. The threat in each category relates only to the health of the command or HSS issues. Similarly, the type of threat can vary greatly with the type of mission or operation (offense, defense, stability, and civil support operations). These categories are—
>
>   (1) Friendly and enemy course of action. Include friendly courses of action best supported from an HSS standpoint; friendly HSS courses of action that best support the mission; and probable enemy courses of action that could affect friendly HSS units, resources, and services.
>
>   (2) Geographic-related threat issues. Include climatic and weather-related threats and their impact on the need for and delivery of HSS and terrain-related issues (best depicted by creating a modified combined-obstacles overlay).
>
>   b. Additional elements of medical information and intelligence can be consolidated into formats that are user-friendly and are available for future planning or other possible contingencies. Databases are particularly useful for managing general information.

Figure F-1. Medical intelligence preparation of the battlefield template (continued)

MEDICAL INTELLIGENCE SUPPORT APPENDIX TO AN INTELLIGENCE ANNEX

F-6. AHS planners develop a medical intelligence support appendix (Figure F-2, pages F-7 and F-8) to the intelligence annex of an OPLAN to facilitate the collection and dissemination of medical information and intelligence and to assure the commander that medical-specific EEFI are addressed. The example shown in Figure F-2 is for a Chairman of the Joint Chiefs of Staff OPLAN. When developing OPLANs/OPORDs at the ASCC level and below, the medical intelligence appendix will be placed under Annex I (Service and Support) in numerical order for ease of use by personnel. Not all OPLANs/OPORDs will require this appendix.

Health Threat and Medical Intelligence

CLASSIFICATION

( ) APPENDIX XX TO ANNEX I TO OPERATIONS ORDER XXXX
( ) **MEDICAL INTELLIGENCE SUPPORT**
( ) REFERENCES: SEE ANNEX B, Chairman of the Joint Chiefs of Staff OPLAN 0400-01

1. ( ) GENERAL. Provide general information on the medical intelligence support to the OPLAN.
    a. ( ) **Purpose.** Focus on the detailed medical intelligence needed to plan and execute military operations across the range of military operations. Medical intelligence identifies environmental and disease threats to U.S. forces and the civilian and military health care capability, infrastructure, and installations of military significance.
    b. ( ) **Relationships.** Specify relationships between the intelligence staff and the AHS, operations, CA, and SO staffs to ensure effective coordination of requirements, priorities, and intelligence.
2. ( ) **MISSION.** Ensure effective coordination between the intelligence staff and the AHS, operations, CA, and SO staffs as a minimum for the use and application of medical intelligence.
3. ( ) **MEDICAL INTELLIGENCE ESTIMATES.** Provide, develop, or obtain estimates on the following:
    a. ( ) **Diseases of Operational Importance in the AO.** Identify disease risks likely to affect U.S. military personnel in the potential AO. Identify variations in the disease situation associated with geography and climate that can be expected throughout the projected deployment period. Identify the disease situation of the population or subpopulations in the potential AO that might influence AHS and CA planning. Identify the naturally occurring infectious diseases within the area that could mask or confine detection or identification of weapons of mass destruction use (biological warfare and chemical warfare agents).
    b. ( ) **Environmental Health Factors of Operational Importance.** Identify the environmental characteristics in the AO that could have an impact on the health of U.S. military personnel. Identify the status of public infrastructures, such as piped-water supply, surface-water supply, water-treatment systems, and sewage-treatment systems that could influence the health and well-being of U.S. forces and indigenous populations. Identify the major sources of industrial and agricultural populations. Identify poisonous plants and animals that could be hazardous to U.S. military personnel in a field environment. Identify other environmental factors as they pertain to the health, welfare, and the specific type of mission of U.S. forces, such as toxic waste dump sites and toxic industrial chemical sources.
    c. ( ) **Civilian Health Care Infrastructure.** Identify the status of the health care infrastructure in the AO. Identify the location, operational status, and capabilities of major medical treatment facilities and other health care-related installations or services. For example, identify clinics, private practices, laboratories, and mental health facilities, as well as the availability of MEDEVAC and transport platforms and services and the availability of medical equipment and medical equipment repair and maintenance. Identify the capabilities and status of health care personnel categories and their relative ability to sustain AHS operations during a national crisis or war. Identify the major pharmaceutical and medical equipment manufacturing plants and their operational capabilities or status. Characterize the system for the supply of blood and blood products. Characterize the blood supply situation, such as availability, collection, and testing.

CLASSIFICATION

Figure F-2. Medical intelligence support appendix format

Appendix F

CLASSIFICATION

    d. ( ) **Military Health Care Infrastructure.** Identify the location, capabilities, and operational status of the military AHS infrastructure. Identify the major military medical treatment facilities, blood banks, research laboratories, and medical supply and logistics depots. Characterize the MEDEVAC system, methodology, and vulnerabilities associated with the system. Identify the casualty mix experienced by enemy forces. Identify and characterize the blood banking and blood supply system. Identify the medical logistics or resupply system. Characterize the ability of the local forces to sustain themselves medically throughout the range of military operations. Identify the capability for medical defense and treatment of local forces and the anticipated compliance with the provisions of the Geneva Conventions by hostile forces.

4. ( ) **FEEDBACK.** Provide feedback and intelligence reporting on medical EEFI using normal intelligence information and reporting procedures as set forth in Annex B.

CLASSIFICATION

Figure F-2. Medical intelligence support appendix format (continued)

Appendix G
Considerations in Planning Medical Evacuations

ARSOF do not have an organic MEDEVAC system. They are dependent upon the Army MEDEVAC system for this support. ARSOF do have an organic capability to effect CASEVAC using ARSOF airframes (those used for infiltration and extraction of ARSOF personnel). During CASEVAC, the casualty may not receive en route medical care unless specific planning and coordination occur to staff the airframe with medically trained personnel before executing CASEVAC operations. Information on the U.S. Army MEDEVAC system is provided in FM 4-02.2, *Medical Evacuation*; FM 8-10-6, *Medical Evacuation in a Theater of Operations Tactics, Techniques, and Procedures*; and JP 4-02.

MEDICAL EVACUATION PLANNING FACTORS

G-1. Planning effective MEDEVAC for ARSOF requires an understanding of ARSOF missions and units. Planning for ARSOF MEDEVAC may differ from standard AHS planning in the following areas:
- Inability to assign dedicated MEDEVAC platforms to all the teams and small units that are often widely dispersed throughout the AO or that are in hostile or denied territory.
- Lack of USAF-approved airfields in many locations in which ARSOF must operate.
- Security requirements of some missions.
- Accountability of sensitive equipment carried by some ARSOF Soldiers. If the ARSOF Soldier is ambulatory, he retains responsibility for any sensitive equipment he has in his possession. If he is unconscious, the equipment is turned over to a team member accompanying the patient. However, if another team member does not accompany the patient, the equipment must be secured until it can be transferred back to the parent unit.
- Individually tailored evacuation plans are required to support numerous small teams deployed to separate locations.

G-2. Although ARSOF have no dedicated assets for MEDEVAC, ARSOF AHS planners must be able to plan and coordinate an efficient chain of CASEVAC from isolated locations anywhere in the world. This evacuation chain requires identification of all specific military assets to complete the mission. AHS planners must then coordinate POCs and every link of the evacuation down to the SOF user level. If required, coordination should include the POC for medical regulating. Once the ARSOF Soldier enters the conventional AHS, medical regulating officers assigned to the medical group or brigade provide medical regulating support. ARSOF AHS planners and ARSOF surgeons must rapidly tailor a CASEVAC plan for ARSOF missions or operations. If ARSOF are assigned to a JSOTF, an ARSOF surgeon, if designated as the JSOTF surgeon, plans for MEDEVAC of the joint forces. ARSOF AHS planners should plan for MEDEVAC in two distinct phases: intratheater and intertheater.

INTRATHEATER EVACUATION

G-3. Within the theater, ARSOF casualties are often evacuated on the aircraft responsible for extracting the other members of the team. Prolonged exfiltration routes in blacked-out aircraft over hostile or denied territory make in-flight patient care delivery difficult. These extraction aircraft must be able to effect rapid communications with the appropriate medical units upon entry into airspace under U.S., allied, or coalition control.

Appendix G

G-4. Coordination for dedicated MEDEVAC platforms must occur to meet the incoming aircraft to evacuate the patient to the appropriate echelon of care. The ARSOF team should retain sensitive equipment and documents in the patient's possession. The team should not transfer these items to the evacuation platform. Because of the classified nature of many ARSOF missions, segregation of the ARSOF patient from other conventional patients may be necessary to protect classified mission parameters from compromise.

INTERTHEATER EVACUATION

G-5. Once the patient enters the conventional AHS, the ARSOF surgeon continues to track ARSOF Soldiers being evacuated out of theater to keep the ARSOF chain of command informed and to ensure that security concerns are addressed. The Joint Patient Tracking Application is a Web-based patient tracking and management tool that collects and reports data on patients arriving at medical treatment facilities from forward-deployed locations. The Joint Patient Tracking Application provides information about the transportation, treatment, and disposition of patients. The referenced Web site is at the following link: *https://jhp.osd.mil/registration.jsp*. Additionally, the USSOCOM Care Coalition liaison officers are available for real-time online viewing by accessing the secure USSOCOM Web site at the following link: *https://sofnet.socom.smil.mil/sites/socs-fc/default.aspx*. This particular Web site requires a request for access and an access code.

G-6. ARSOF AHS planners must continuously apprise the situation to ensure that plans remain sufficiently flexible to provide the necessary support when it is required. They must also maintain active liaison with the conventional AHS units that will provide the support once the ARSOF patient is extracted from the AO. Planners should—

- Determine the airfield to which ARSOF patients will be evacuated.
- Determine if any medical equipment and supplies are required to sustain the ARSOF patient while in-flight.
- Coordinate for the augmentation of medically trained personnel to be aboard the airframe when the ARSOF patient is picked up. This asset may come from organic ARSOF personnel because of the classified parameters of the mission.
- Coordinate for dedicated MEDEVAC support to be present at the destination airfield. ARSOF patients extracted from hostile or denied territory are normally taken to a medical treatment facility for evaluation and stabilization before further evacuation. ARSOF patients are not normally evacuated directly to a mobile aeromedical staging facility, as these facilities are not staffed or equipped to provide stabilizing medical care.

EVACUATION FROM HOSTILE OR DENIED TERRITORY

G-7. ARSOF planners and SFODA commanders must develop tentative plans for the evacuation of ARSOF patients from hostile or denied territory, when feasible. Planners must consider all options that will not compromise the security of the operation. Conventional MEDEVAC platforms cannot normally provide support while ARSOF teams are deployed.

G-8. ARSOF planners should consider the following factors:

- Classified nature of the mission and the probable outcome if compromised.
- Availability of opportune vehicles and aircraft, such as resupply platforms.
- Availability of HN transportation resources, such as pack animals or civilian transportation assets.
- Infiltration and exfiltration routes.
- Requirements for special medical equipment and supplies.
- Availability of HN medical care facilities, equipment, and supplies to stabilize the patient for an arduous ground evacuation.
- Probable weather in the AO. Reduced visibility may enhance the chance of successfully exfiltrating the patient, or inclement weather such as snow or extremely cold temperatures may impose special requirements for sustaining the patient until he can be evacuated.
- Landing area requirements and the maximum time the airframe can loiter while awaiting pickup of the patient.

Glossary

SECTION I – ACRONYMS AND ABBREVIATIONS

A	airborne
ACSA	acquisition and cross-servicing agreement
A/DACG	arrival/departure airfield control group
ADP	Army doctrine publication
AHS	Army Health System
ALE	Army special operations forces liaison element
ALOC	air line of communications
AO	area of operations
AOC	area of concentration
AOI	area of interest
AOR	area of responsibility
AR	Army regulation
ARNG	Army National Guard
ARSOAC	Army Special Operations Aviation Command
ARSOF	Army special operations forces
ASA(ALT)	Assistant Secretary of the Army for Acquisition, Logistics, and Technology
ASCC	Army Service component command
ASL	authorized stockage list
ASPO	Army special operations forces support operations
ATP	Army techniques publication
ATTP	Army tactics, techniques, and procedures
BSC	battalion support company
CA	Civil Affairs
CAIS	civil authority information support
CAISI	Combat-Service-Support Automated Information Systems Interface
CAO	Civil Affairs operations
CASEVAC	casualty evacuation
CBRN	chemical, biological, radiological, and nuclear
CCDR	combatant commander
CCO	contingency contracting officer
CERP	Commanders' Emergency Response Program
CJA	Command Judge Advocate
C-MNS	Combat-Mission Needs Statement
CMO	civil-military operations
Comm	commercial
COMSEC	communications security

Glossary

CONPLAN	concept plan	
CONUS	continental United States	
COSC	combat and operational stress control	
CSSB	combat sustainment support battalion	
CTA	common table of allowance	
CUL	common-user logistics	
DA	Department of the Army	
DCS	deputy chief of staff	
DCSAC	Deputy Chief of Staff, Acquisition and Contracting	
DOD	Department of Defense	
DOS	Department of State	
DSN	Defense Switched Network	
DSOR	deployment statement of requirement	
ECC	Expeditionary Contracting Command	
ECOP	equipment common operational picture	
EEFI	essential elements of friendly information	
ESC	expeditionary sustainment command	
ESD	equipment sourcing document	
FARP	forward arming and refueling point	
FHP	force health protection	
FID	foreign internal defense	
FM	field manual	
G-3	Deputy Chief of Staff for Operations and Plans	
G-4	Deputy Chief of Staff for Logistics	
G-8	Deputy Chief of Staff for Resource Management	
GCC	geographic combatant commander	
GSB	group support battalion	
HEMTT	Heavy Expanded Mobility Tactical Truck	
HHC	headquarters and headquarters company	
HN	host nation	
HNS	host-nation support	
HQ	headquarters	
HQDA	Headquarters, Department of the Army	
HSOC	home station operations center	
HSS	health service support	
IAW	in accordance with	
INSCOM	United States Army Intelligence and Security Command	
ISB	intermediate staging base	
J-3	operations directorate of a joint staff	
J-4	logistics directorate of a joint staff	

Glossary

J-6	command, control, communications, and computer systems directorate of a joint staff
JCET	joint combined exchange training
JCRM	Joint Capability Requirements Manager
JOA	joint operations area
JOS	joint operational stocks
JP	joint publication
JSOTF	joint special operations task force
JTF	joint task force
LCOP	logistical common operating picture
LSC	Lead Service for Contracting
MEDEVAC	medical evacuation
METT-TC	mission, enemy, terrain and weather, troops and support available, time available, and civil considerations
MFP	Major Force Program
MHE	materials handling equipment
MIPB	medical intelligence preparation of the battlefield
MIS	Military Information Support
MISG(A)	Military Information Support group (airborne)
MISO	Military Information Support operations
MISOC	Military Information Support Operations Command
MOA	memorandum of agreement
MOS	military occupational specialty
MPA	Military Personnel, Army
MRE	meal, ready to eat
MST	maintenance support team
MTF	medical treatment facility
MTOE	modified table of organization and equipment
NATO	North Atlantic Treaty Organization
NCO	noncommissioned officer
NDAA	National Defense Authorization Act
NSN	National Stock Number
O&M	operation and maintenance
OCONUS	outside the continental United States
OMA	Operations and Maintenance, Army
ONS	operational needs statement
OPFUND	operational fund
OPLAN	operation plan
OPORD	operation order
OTSG	Office of the Surgeon General
PARC	principal assistant responsible for contracting

PBO	property book officer	
POC	point of contact	
POL	petroleum, oils, and lubricants	
RDT&E	research, development, test, and evaluation	
RM	resource management	
RSC	Ranger Support Company	
RSOD	Ranger Support Operations Detachment	
RSOI	reception, staging, onward movement, and integration	
S-1	personnel officer	
S-2	intelligence officer	
S-3	operations and training officer	
S-4	logistics officer	
SACEUR	Supreme Allied Commander, Europe	
SAMS-E	Standard Army Maintenance System-Enhanced	
SARSS	Standard Army Retail Supply System	
SASMO	Sustainment Automation Support Management Office	
SB(SO)(A)	Sustainment Brigade (Special Operations) (Airborne)	
SF	Special Forces	
SFG	Special Forces group	
SFODA	Special Forces operational detachment A	
SFODB	Special Forces operational detachment B	
SO	special operations	
SOA	special operations aviation	
SOAL	special operations acquisition and logistics	
SOAR	special operations aviation regiment	
SOCCENT	Special Operations Component, United States Central Command	
SOCEUR	Special Operations Component, United States European Command	
SOCM	special operations combat medic	
SOF	special operations forces	
SOFORGEN	Special Operations Force Generation	
SOFSA	special operations forces support activity	
SOP	standing operating procedure	
SOR	statement of requirement	
SOTF	special operations task force	
SPO	support operations	
SPTCEN	support center	
SSA	supply support activity	
SSAVIE	Special Operations Forces Sustainment, Asset Visibility, and Information Exchange	
SSSC	self-service supply center	
STAMMIS	standard Army multi-command management information system	

Glossary

STU	secure telephone unit
TMDE	test, measurement, and diagnostic equipment
TSC	theater sustainment command
TSCP	theater security cooperation plan
TSOC	theater special operations command
TTP	tactics, techniques, and procedures
UMT	unit ministry team
U.S.	United States
USAF	United States Air Force
USAFRICOM	United States Africa Command
USAJFKSWCS	United States Army John F. Kennedy Special Warfare Center and School
USAMC	United States Army Materiel Command
USAR	United States Army Reserve
USARCENT	United States Army, Central Command
USARPAC	United States Army, Pacific Command
USARSO	United States Army, Southern Command
USASOC	United States Army Special Operations Command
USC	United States Code
USCENTCOM	United States Central Command
USEUCOM	United States European Command
USPACOM	United States Pacific Command
USSOCOM	United States Special Operations Command
USSOUTHCOM	United States Southern Command
USTRANSCOM	United States Transportation Command
UW	unconventional warfare

This page intentionally left blank.

References

REQUIRED PUBLICATIONS
These documents must be available to intended users of this publication.

None.

RELATED PUBLICATIONS
These documents contain relevant supplemental information.

ARMY FORMS
Department of the Army Forms are available on the Army Publishing Directorate Web site (www.apd.army.mil).

DA Form 285. *Technical Report of U.S. Army Ground Accident.*
DA Form 348. *Equipment Operator's Qualification Record (Except Aircraft).*
DA Form 2028. *Recommended Changes to Publications and Blank Forms.*
DA Form 3953. *Purchase Request and Commitment.*

ARMY PUBLICATIONS
ADP 3-0. *Unified Land Operations.* 10 October 2011.
ADRP 4-0. *Sustainment.* 31 July 2012.
AR 15-6. *Procedures for Investigating Officers and Boards of Officers.* 2 October 2006.
AR 27-10. *Military Justice.* 3 October 2011.
AR 27-20. *Claims.* 8 February 2008.
AR 40-68. *Clinical Quality Management.* 26 February 2004.
AR 71-32. *Force Development and Documentation – Consolidated Policies.* 3 March 1997.
AR 115-11. *Geospatial Information and Services.* 10 December 2001.
AR 165-1. *Army Chaplain Corps Activities.* 3 December 2009.
AR 710-2. *Supply Policy Below the National Level.* 28 March 2008.
AR 715-9. *Operational Contract Support Planning and Management.* 20 June 2011.
ATP 4-16. *Movement Control.* 5 April 2013.
ATTP 1-19. *U.S. Army Bands.* 7 July 2010.
ATTP 4-02. *Army Health System.* 7 October 2011.
ATTP 4-10. *Operational Contract Support Tactics, Techniques, and Procedures.* 20 June 2011.
ATTP 4-33. *Maintenance Operations.* 18 March 2011.
FM 1-0. *Human Resources Support.* 6 April 2010.
FM 1-04. *Legal Support to the Operational Army.* 18 March 2013.
FM 1-05. *Religious Support.* 5 October 2012.
FM 1-06. *Financial Management Operations.* 4 April 2011.
FM 3-05.160. *Army Special Operations Forces Communications System.* 15 October 2009.
FM 3-34.400. *General Engineering.* 9 December 2008.
FM 3-53. *Military Information Support Operations.* 4 January 2013.
FM 3-57. *Civil Affairs Operations.* 31 October 2011.
FM 3-76. *Special Operations Aviation.* 28 October 2011.

References

FM 4-02.2. *Medical Evacuation.* 8 May 2007.

FM 4-02.12. *Army Health System Command and Control Organizations.* 26 May 2010.

FM 4-02.43. *Force Health Protection Support for Army Special Operations Forces.* 27 November 2006.

FM 8-10-6. *Medical Evacuation in a Theater of Operations Tactics, Techniques, and Procedures.* 14 April 2000.

FM 27-10. *The Law of Land Warfare.* 18 July 1956.

TC 18-01. *Special Forces Unconventional Warfare.* 28 January 2011.

USASOC Policy No. 32-09. *Operational Funds (OPFUNDS).* 3 September 2009.

USASOC Regulation 350-1. *ARSOF Active Component and Reserve Component Training.* 5 April 2011.

DEPARTMENT OF DEFENSE PUBLICATIONS

DD Form 93. *Record of Emergency Data.*

DD Form 200. *Financial Liability Investigation of Property Loss.*

DD Form 1348-1A. *Issue Release/Receipt Document.*

DOD Financial Management Regulation 7000.14-R, Volume 12. *Special Accounts Funds and Programs.* 25 January 2012.

DOD Instruction 7250.13. *Use of Appropriated Funds for Official Representation Purposes.* 30 June 2009.

JOINT PUBLICATIONS

JP 4-0. *Joint Logistics.* 18 July 2008.

JP 4-02. *Health Service Support.* 26 July 2012.

JP 4-10. *Operational Contract Support.* 17 October 2008.

OTHER PUBLICATIONS

AF IMT Form 3822. *Landing Zone Survey.*

AF IMT Form 3823. *Drop Zone Survey.*

National Defense Authorization Act, Section 1206.

National Defense Authorization Act, Section 1208.

SGLV 8286. *Servicemembers' Group Life Insurance Election and Certificate.*

Standard Form 44. *Purchase Order-Invoice-Voucher.*

Standard Form 1402. *Certificate of Appointment.*

Uniform Code of Military Justice, Article 32. *Investigation.*

USC, Title 10. *Armed Forces.* 3 January 2012.

USC, Title 10, Section 127. *Emergency and Extraordinary Expenses.* 3 January 2012.

USC, Title 10, Section 127b. *Assistance in Combating Terrorism: Rewards.* 3 January 2012.

USC, Title 10, Section 2341. *Authority to Acquire Logistic Support, Supplies, and Services for Elements of the Armed Forces Deployed Outside the United States.* 3 January 2012.

USC, Title 10, Section 2342. *Cross-Servicing Agreements.* 3 January 2012.

USC, Title 22, Section 2778. *Control of Arms Exports and Imports.* 3 January 2012.

Index

A

Army Health System (AHS), vi, 1-4, 3-6, 3-8, 4-6, 4-9, 5-4, 5-11, 7-5, 8-6, 9-1, 9-4, 9-9, 9-10, A-3, F-1, F-2, F-7, G-1, G-2

Army Service component command (ASCC), 1-1, 1-2, 1-4, 1-5, 2-1, 2-2, 2-5 through 2-11, 3-1 through 3-5, 4-1, 4-8, 5-2, 5-11, 6-3, 6-6, 6-9, 6-10, 7-1, 7-4, 7-5, 8-3 through 8-5, 11-3, E-1, F-7

Army special operations forces liaison element (ALE), vi, 1-2, 2-2, 2-5, 2-9, 3-1 through 3-6, 4-2, 5-10, 6-4, 6-6 through 6-8, 6-10, 7-1, 7-3 through 7-5, 8-3 through 8-5, E-1

Army special operations forces support operations (ASPO) cell, iv, 1-2, 3-3, 3-7, 7-4, 8-3

C

casualty evacuation (CASEVAC), 4-10, 5-11, 9-4, G-1

chemical, biological, radiological, and nuclear (CBRN), 5-5, 5-10, E-3, E-5, F-1, F-2

Civil Affairs (CA), v, vi, 1-1, 1-3, 2-2, 2-3, 7-1, 7-3 through 7-5, 9-3, 9-4, 9-7

contingency planning, 6-3, 6-4, 7-4, 8-2, 8-5

crisis action planning, 2-10, 6-3, 7-4, 7-5, 8-5

F

field services, 1-8, 5-7

forward support company, 3-3, 3-8, 4-6, 4-8

G

group support battalion (GSB), 1-2, 1-3, 2-2, 2-5, 3-3, 3-6, 4-1 through 4-9, 4-11, 4-18, 6-4, 7-4

H

host-nation support (HNS), 1-6, 1-10, 2-6 through 2-9, 4-1, 4-4, 4-17, 6-3, 6-6, 6-10, 11-4, A-2, A-6, E-1

J

joint special operations task force (JSOTF), 1-1 through 1-3, 2-2, 2-5, 2-9, 2-10, 3-1, 3-3, 3-4, 3-6, 4-1 through 4-5, 4-8, 6-4, 6-6, 6-7, 7-1, 7-4, D-2, E-6, G-1

M

medical evacuation (MEDEVAC), 4-8, 4-10, 4-12, 5-11, 9-4, 9-7 through 9-9, A-6, D-5, E-10, G-1, G-2

Military Information Support operations (MISO), v, 1-1, 1-3, 8-1 through 8-6

P

patient tracking, G-2

R

Ranger Special Troops Battalion, 1-3, 5-1, 5-2, 5-11

Ranger support company (RSC), vi, 1-3, 5-1, 5-2, 5-5 through 5-7, 5-9, 5-10, 7-4

S

Special Forces
operational detachment A (SFODA), 1-2
operational detachment B (SFODB), 4-10

special operations combat medic (SOCM), 6-7, 6-8, 9-3, 9-4

statement of requirement (SOR), 2-8, 4-4, 6-3, 6-5 through 6-7, 7-4, 7-5, 8-2, 8-3, 8-5, E-1

sustainment brigade, v, vi, 1-2, 1-5, 2-1, 2-2, 3-1, 3-3, 3-5, 6-9

T

theater opening, 4-2

theater special operations command (TSOC), vi, 1-2, 1-4, 1-10, 1-11, 2-2, 2-3, 2-5, 2-7 through 2-9, 2-11, 3-1, 3-2, 3-4, 3-5, 3-8, 4-8, 4-17, 4-18, 6-1, 6-4, 6-7, 7-1, 7-3 through 7-5, 8-1, 8-3 through 8-5, 9-2, 9-7, 11-2, B-1, B-3

theater sustainment, 4-4, E-1

theater sustainment command (TSC), vi, 1-3, 1-5, 2-1, 2-9 through 2-11, 3-1, 3-3, 3-6, 4-2, 4-3, 4-5, 4-7, 5-6, 5-10, 5-11, 6-4 through 6-7, 6-10, 7-1, 7-3, 7-5, 8-5

W

warfighting function, vi, 1-1, 1-7, 1-8, 4-1, 9-1

This page intentionally left blank.

ATP 3-05.40 (FM 3-05.140)
3 May 2013

By Order of the Secretary of the Army:

RAYMOND T. ODIERNO
General, United States Army
Chief of Staff

Official:

JOYCE E. MORROW
Administrative Assistant to the
Secretary of the Army
1300701

DISTRIBUTION:
Active Army, Army National Guard, and United States Army Reserve: Not to be distributed; electronic media only.

PIN: 103306-000

www.ingramcontent.com/pod-product-compliance
Lightning Source LLC
Chambersburg PA
CBHW071208240526
45470CB00018B/1600